Safety in Petroleum Industries

Safety in Petroleum Industries

Dhananjoy Ghosh

CRC Press
Taylor & Francis Group
Boca Raton London New York

CRC Press is an imprint of the
Taylor & Francis Group, an **informa** business

First edition published 2021
by CRC Press
6000 Broken Sound Parkway NW, Suite 300, Boca Raton, FL 33487-2742

and by CRC Press
2 Park Square, Milton Park, Abingdon, Oxon, OX14 4RN

Library of Congress Cataloging-in-Publication Data

ISBN: 978-0-367-65384-2 (hbk)
ISBN: 978-0-367-65393-4 (pbk)
ISBN: 978-1-003-12925-7 (ebk)

Typeset in Palatino
by MPS Limited, Dehradun

This book is dedicated to the memory of my heavenly parents.

Contents

Acknowledgments

Some of my colleagues in the industries and academia, who had seen me handling the practical problems of all refinery technologies and maintenance, and simultaneously attending forums in the industry and universities to provide demonstrations to the engineers, were vocal in their suggestions that I write a book on safety in petroleum industries. This was a subject of most serious concern in the industry due to the prevalence of occasional unsafe incidents, even after implementation of advanced technologies in the field. Their encouragement made me consider putting efforts in this direction. Today I am very happy to make it possible and convey my thanks and gratitude to all my well-wishers.

I especially acknowledge the guidance of Dr. Uttam Ray Chaudhuri, former professor of Calcutta University, India, for offering his views on the pros and cons of publication, and for providing some principal thoughts in authoring the book.

My wife, Mita, has observed me for more than 35 years and obviously well understands about my aptitude on handling different situations in my profession. She also asked me to write a book on safety. Today, I am very happy to acknowledge her deep inspiration in writing this book.

Preface

While demonstrating his knowledge to engineers in the industries and students in universities, the author realized that the book should be simple in expression as well as very practical in its focus on the application of knowledge. The intention is that the book is user-friendly to chemical engineers and operators as well as engineering students, thereby contributing rightfully to enhance safety and sustenance of industrial operations, a matter of concern in the industries, due to the level of accidents continuing to occur in this field.

A number of chapters on safety have been included in such a way that not only petroleum industries but also other process industries should benefit by using the book for study and reference. Attention has been given to cover the chapters in a way that the readers should find the book highly relevant in day-to-day application in the industries, as well as in the academia to upgrade the skills of the students.

The book covers process safety management, the technology selection process, hazard identification, checklisting, risk analysis and assessment calculation, HAZOP studies, engineering standards, including electrical and instrumentation, plant operation, maintenance and inspection, fire fighting, health and hygiene, project safety, safety performance indicators, safety reporting, safety training, safety audits, international guidelines on occupational health and safety (OHS), safety tips for operation and a number of accident case studies in the petroleum industries.

The chapters/subchapters are focused on petroleum as well as chemical process industries, though some of the above-mentioned processes mainly focus on onshore petroleum industries and particularly petroleum refineries. However, the book should also benefit offshore petroleum installations and all chemical industries except in the matter of specific safety features unique to them.

The readers, if they go through the specific chapters that include checklisting, safety tips for operation, and all case studies, and keep these as priorities for application, will find it very comfortable to sustain the operations safely and uninterrupted, when assigned to such areas.

Author

Dhananjoy Ghosh is a chemical engineer, an industrial professional, and visiting faculty in various engineering universities. He graduated with honours in chemistry and earned his PhD in petroleum technology from the University of Calcutta, India. Since the beginning, his focus has been on innovation, troubleshooting and safety. He has worked in many refineries to assist with the safe commissioning of various projects. Due to his expertise on innovation, he developed new technology on a process, "Re-extraction of aromatics from petroleum residual extract" in 1995 and received a national award from NPMP (under the ministry of India) in 2000. In 1996, he developed another technology, "Simultaneous de-oiling & de-waxing in a single unit to produce quality Base Oil & Micro-crystalline Wax simultaneously" with the objective of integration and energy savings, which was commercially implemented in 2002. He has many research publications in both national and international journals. He has served petroleum refineries for 35 years and chemical and petrochemical industries as a head, though for a shorter duration. He also has experience in working as an expert panel member for the audit of Indian refineries for the benchmarking of their performance.

1

Preliminaries

1.1 Concept and Importance of Safety

There is always a safe way to do every job—accidents don't just happen. Rather, they are caused. To prevent accidents, it is vital to teach safety to all new hands. To begin, let's review the concept of safety.

The importance of safety practices in the petroleum and chemical process industries are of utmost importance due to the potential hazard to human lives and plant assets. In fact, the sustainability of an industry's operations for all stakeholders depends on the extent to which it is operated without injury and fatality, as well as without damage to plant assets to ensure uninterrupted operation.

A major concern of safety in industrial plants is the occurrence of fires and explosions. To emphasize the importance of this issue, industrial safety is highlighted in the form of codes, standards, technical papers and engineering design.

Occupational safety and health administration (OSHA) publication 3073 defines a hazardous location as follows:

Hazardous locations are areas where flammable liquids, gases or vapors and combustible dust exist in sufficient quantities to produce an explosion or fire. In hazardous locations, specially designed equipment and special installation techniques must be used to protect against the explosive and inflammable potential of these hazardous substances.

1.2 Definition of Explosion and Its Relevance

An explosion is the sudden release of pressure that causes a rapid and large volume expansion of flammable liquids, gases or vapors due to fire. Sometimes explosions without fire also can take place if the liquids, gases or vapors are below their ignition temperature or flash point. As a rule, three basic requirements must be met for an explosion to take place in atmospheric air:

- A flammable substance needs be present in substantial quantity to produce an ignitable or explosive mixture.
- Oxygen must be present in sufficient quantity with the inflammable substance to produce the explosive mixture.
- A spark or high heat/temperature must exist.

Note:

i. The above three conditions form the 'fire triangle,' as it is commonly called.
ii. Every specific gas or vapor has its upper and lower flammability limit. If the concentration of the substance in oxygen or air is either above or below that specific limit, ignition might occur but flames will not propagate. Thus, there will be no explosion.

1.3 Safety in Engineering

In view of the above, international organizations have introduced safety codes that are followed by the industries to establish a standardized, safe operation. In addition, respective countries have established statutory bodies to ensure implementation of international safety standards and/or set up their own guidelines of safety standards to be followed by the respective industries.

In line with this objective, the subject of safety in engineering has acquired importance in engineering program degrees and course curriculums and has been structured into multiple sub-subjects so it is inculcated as fundamental knowledge in engineering.

In the mechanical engineering industries, safety issues are subcategorized as follows:

1. Principles of machine guarding.
2. Safety in welding and gas cutting.
3. Selection and Suitability.
4. Safety in cold forming and hot working of metals.
5. Safety in inspection and testing.
6. Handling and storage of materials.

7. Mechanized handling of materials.
8. Working at height.
9. Use of tools.
10. Safety in the workplace.
11. Plant design and housekeeping.

All the above-mentioned topics are elaborated below as exercises in mechanical engineering in any industry in day-to-day activities.

- *Principle of machine guarding:*
  Guarding during maintenance, Zero Mechanical State (ZMS), operation of protective devices, machine guarding and its types (such as fixed guard, interlock guard, automatic guard, trip guard, electron eye, positional control guard, fixed guard fencing, guard construction, guard opening, etc.), along with benefits of guarding systems.
- *Selection and suitability:*
  Selection of various machines, such as the lathe, drilling, boring, milling, grinding, shaping, sawing, shearing presses, forge-hammer, flywheels, shafts, couplings, gears, sprockets, wheels, chains-pulleys, belts and the criteria for authorized entry to hazardous installations.
- *Safety in welding and gas cutting:*
  Gas welding and cutting, resistance welding, arc welding and cutting, personal protective equipment (PPE) training, safety precautions in brazing, soldering and metalizing, explosive welding, selection and maintenance of the associated equipment and instruments, distribution and handling of industrial gases, color coding, flashback arrestor, leak detection, pipe line safety and storage/handling of gas cylinders, etc.
- *Safety in cold forming and hot working of metals:*
  Cold working, power presses, auxiliary mechanisms, feeding and cutting mechanism, hand or foot-operated presses, power press electric controls, power press set up and die removal, inspection and maintenance, metal sheers, press brakes and hot working safety in forging, hot rolling mill operation, safe guards in hot rolling mills, bending of pipes, hazards and control measures, safety in gas furnace operation, cupola, crucibles, ovens, foundry health hazards, work environment, material handling in foundries, foundry production cleaning, finishing foundry processes, etc.
- *Safety in finishing, inspection and testing:*
  Heat treatment operations, electro plating, paint shops, sand and shot blasting, inspection and testing like dye penetration, radiography and hydro testing, dynamic balancing, hydro testing,

valves, boiler drums and headers, pressure vessels, air leak test, steam testing, radiation hazards, Indian Boilers Regulations, etc.

- *Manual handling and storage of materials:*
  Basic principle of correct lifting and handling of materials; avoidance of excessive muscular efforts, maximum loads that may be carried, carrying of objects of different shapes, size and weight, safe use of accessories for manual handling, storage of materials, ergonomics of manual handling and storage, etc.
- *Mechanized handling of materials:*
  Safety aspects in design and construction, testing, use and care, signaling, inspection and maintenance, operation; safe working load, inspection and maintenancc of lifting tackles, competencies of persons, duties and responsibilities under the various legislations.
- *Working at height:*
  Safety features associated with design, construction and use of stairways, runs, ramps, gangways, floors, ladders of different types, scaffoldings of different types, working on roofs, other safety requirements while working at heights, prevention of falls of persons at floor level from height or slipping.
- *Uses of tools-hand tools and portable tools:*
  Main causes of tool accidents, control of tool accidents, purchase, storage and supply of tools, inspection and maintenance of tools, tempering of tools, safe ending and dressing of certain tools like handles, etc., safe use of various types of hand tools for metal cutting, wood cutting, miscellaneous cutting work, material handling, and other hand tools, such as torsion tools, shock tools, non-sparking tools, portable power tools, etc.
- *Safety in the workplace:*
  Workplace design, improving safety and productivity through workplace design, technical and engineering control measures, control measures against human error, preventive maintenance, standards and code of practices for plant and equipment, etc.
- *Plant design and housekeeping:*
  Safety in good housekeeping, typical accidents due to poor housekeeping, disposal of scrap and other trade wastes, prevention of spillage, marking of gangways and use of color codes as an aid for good housekeeping, clean-up campaigns, cleaning methods, inspection, employee responsibilities, check-listing, etc.

In electric power generation and distribution industries, the following additional sections are covered:

- Theories on power generation and application, such as steam generators with coal/oil fired boilers, gas turbines with or without an exhaust heat recovery system, nuclear power generation, hydraulic power generation, wind mills, solar power, etc.
- Distribution systems, along with transformers and substations covering the power control centre (PCC), motor control centre (MCC), electric motors of high tension (HT) or low tension (LT) class, distribution cable networking HT and/or LT, power cabling, control cabling, relay circuits, UPS (uninterrupted power supply), etc.

In civil engineering, safety is applicable everywhere—from mechanical and electrical infrastructure, such as roads, bridge-building, underground construction up to all process industries. The following are covered:

- Construction mix studies and engineering.
- Structural engineering, including structure stability design.
- Seismic engineering.
- Soil engineering.
- Painting.
- Insulation.
- Fire proofing.
- Refractory engineering.

1.4 Safety in Petroleum Industries

1.4.1 Features of Petroleum Industries

A petroleum refinery and its downstream industries should be understood as a complex system comprising of many subsections. Petroleum refineries have many process plants: the crude distillation unit, vacuum distillation unit, naphtha hydrotreatment and reforming unit, diesel hydrotreatment, hydrodesulfurization unit, hydrocracker unit, fluid catalytic cracking (FCC) unit, Alkylation unit, Isomerization unit, bitumen unit, visbreaking or delayed coker unit (DCU), deep catalytic cracking (DCC) unit, Lube units, amine absorber units, sour gas treatment unit, sulphur recovery unit, flare section, captive power plant, utilities generation unit, effluent treatment plant (ETP), and producing various products starting from LPG, propane, butane, naphtha, Motor Spirit (MS/Gasoline), kerosene, ATF, High Speed Diesel (HSD), bitumen, furnace oil, lubes and waxes, in addition to various speciality products, such as hexane, low aromatic solvent, benzene, toluene, ethylene, propylene, etc. To ensure smooth storage and dispatch of these products, the refineries also maintain large numbers of storage tanks, along with various dispatch facilities, such as road tankers, railway wagon loading, pipeline transfer, ship loading and coke dispatch facility, as well as a crude oil receiving facility. Petrochemical plants also contain similar facilities but the process plants are different. They include a naphtha or gas cracker unit, ethylene and propylene recovery unit, benzene, toluene and para-xylene recovery units, MTBE unit, and various polymer units, such as poly ethylene, poly propylene, etc. Also, instead of crude, their raw material is naphtha/ LPG/NG, and receiving systems are different. Their main products dispatched are in solid form, such as poly ethylene and poly propylene for which there are bagging, storing and dispatch facilities available. Both refineries and petrochemical plants have to maintain a large administration, procurement department, fire and safety department, marketing department, quality control department, technical service department, engineering service department, public relation department, legal department, estate department, and so on. In the upstream the oil fields have different technologies in operation—whether it is on shore or off shore—but they also have crude oil and gas separation units in on-shore like refineries but, unlike with refineries, processes are simple. However, they also have to maintain different supporting departments, as mentioned above.

1.4.2 Safety Aspects in Petroleum Industries

Petroleum products are highly flammable. Some of the liquids flowing in the pipelines are at temperature above the flash point of the liquids. Electrically-operated higher capacity pumps and compressors, handling flammable hydrocarbon, run continuously. For maintaining various operating parameters different kinds of valves in process pipelines are required. Many toxic by-products, such as SO_2, CO_2, H_2S, etc., are generated during hydrocarbon processing. Many automated control systems are required to be in place in the processes, not only for uninterrupted production and quality, but also to achieve integrated safety to avoid failure and hazard. There is also hazard in storage tanks and vessels containing highly flammable liquids. Various projects activities are always carried out in the complex, including various equipment handling, major constructions, etc. During shutdown, equipment such as vessels, columns, exchangers and furnaces are opened and maintenance activities such as confined space entry, hot work, working at height, etc., are carried out on a major scale.

From the above, it is evident that to understand safety and manage the risk therein, the entire system has to be divided into various subsystems on a departmental basis (as mentioned above) and safety aspects need to be understood and analyzed accordingly.

2

PSM (Process Safety Management)

PSM describes the process safety management to be adopted in any process industry which is very vulnerable with respect to its size and complexity of the processes. Principally it is divided into two categories; namely, commitment to safety and understanding and managing risk.

2.1 Commitment to Safety

To create a good record of few or no accidents, the industry needs a commitment to safety as follows:

It should have a process safety culture; i.e., combined in all levels in the organization contributes to a culture of safety which is reflected by a good reporting and auditing system established across the organization. It should ensure compliance with safety standards in design and layout, operating practices, maintenance and inspection, environmental protection, safety and fire protection, electrical equipments and facilities, EPC (Engineering and Project Construction), safety program and activities, training and development, and the like. It should ensure workforce involvement which covers guidance in using the knowledge of people close to the process. Some elements in commitment to process safety are as follows:

- Adherence to use of PPE (protective personal equipment).
- Following an approval system.
- Issuing daily/monthly/quarterly/yearly incident reports.
- Starting of work with daily safety talks.
- Starting every meeting with safety talks.
- Conducting safety quiz/competition among the employees.
- Observing and conducting a program on fire and safety weeks.
- Conducting a safety awareness program among the neighbors.
- Conducting mock drills.
- Participating in a national safety competition.

- Establishing risk management and an all-management system with respect to safety parameters.
- Issuing an annual sustainability report in line with IFC guidelines.

2.2 Understanding and Managing Risk

The workforce should understand the risk of a process plant and manage the risk commensurate to the commitment to process safety, as mentioned above. The industry should have developed a process knowledge management system with respect to knowledge of technologies, training need identification, training assessment, well documented procedures on design, operation and maintenance, all data records in place and all incident records and lessons documented. Also, all the documents should be easily traceable. The industry should also have a hazard identification and risk analysis system to strengthen the safety system.

Hence, the industries should build up a process safety management system (PSM) to demonstrate the commitment to safety, and achieve the highest level of risk management practices. There are 14 elements of PSM as follows:

- Employee participation.
- Risk studies and assessment and HAZOP studies.
- Strict adherence to standard operating procedures (SOP).
- Establishment of manuals and follow-up management to ensure safe working practices.
- Asset integrity and reliability.
- Contractor management.
- Training and performance assurance.
- Management of change (MOC).
- Operational readiness.
- Daily operation.
- Incident investigation.
- Emergency management.
- Establishment of all ISO systems, including OSHA.
- Compliance audit.

Risk studies and assessment and HAZOP studies, SOP, manual management, MOC, operational readiness, daily operation and emergency management are discussed separately in the next chapter. The compliance audit, like the safety audit, is also discussed in the next chapter on safety practices.

An asset and integrity management system relates to maintenance and inspection departments, ensuring the uninterrupted availability of rotary and stationary equipment for operation. In earlier days the system was a manual process with respect to documentation, historization, and standardization on quality assurance. In current times software systems driven with data banks from the designer and continuous validation from the operating system have been introduced where integration of online operation DCS data, online quality control data on operation, SAP data on maintenance history and spare inventory and on stores and purchases, finance system online data, technical services online performance evaluation data (wherever allowed), and online tank farm accounting and management data systems are done to a single system to ensure quality asset integrity management. The efficacy of the system depends not only on integration, but also (and mainly) on, strength of corrosion control data bank and documentation and rotary equipment performance histories.

ISO systems today are implemented by all process and engineering organizations. However, many industries are not carrying out term audits and periodical surveillance audits to reactivate the approval of the systems by authorized bodies which is a must for holding the quality systems. Presently, industries are following three systems with respect to quality systems: ISO-9000 for production and design, ISO-14001 for environment control management and OHSAS-18001 for occupational safety and health administration.

Also, ISRS (international safety rating system) is followed by some industries. This is a questionnaire-based system and is very practical in the sense that filling the questions in respective safety elements means accepting the challenge for a surveillance audit by the authorized external approving body, one then obtains numerical scores in each element of safety, and finally receives the international ranking.

Employee participation means employees in production, maintenance and all stakeholders related to production. In this management system, level of participation in various aspects of industrial activities is demonstrated through organogram, committee, subcommittee members selection in all defined classes of activities, as well as, from time to time, specialized ad hoc committees for newly arising situations through an approval system run by competent authorities.

3

Organization Safety Procedures

3.1 Technology Selection

While initiating a venture to open an industry or revamp the industry, safety starts with technology selection after the project objectives are defined. A petroleum industry consists of many processing units and associated auxiliary facilities like utilities, receiving feedstock(s), intermediate, if any, storage and dispatch of final products. The plot plan of the grass root plant or revamp of the plant, though tentatively finalized with preliminary information, is to be subsequently revised based on selection of technologies for the processing unit(s). The combination of the process units selected to operate the industry is called configuration. Initially, the industry has to choose a number of configuration alternatives to identify those best possible to meet the objectives with respect to product market, profitability, investment, safety and environmental impact.

Technology(s) selection for the process unit(s) determines the QRA (Quantitative Risk Assessment) and EIA (Environment Impact Assessment). Hence, technology selection is first and foremost important in opening up the industry or revamping the existing industry. As an example, one typical configuration for an integrated petroleum refinery and petrochemical complex is shown in Figure 3.1.

As seen above, various process technologies like, CDU/VDU, NHT/Platformer, DHT, RDS, FCC, Naphtha Cracker and PX have been selected after studying alternative configurations with respect to profitability and investment. But it does not end with the process of technology(s) selection because the following are to be satisfied before finalizing these technologies:

1. A proven track record of safety should be ensured, along with proven records of operational performance while selecting the engineering design, technology(s) of respective process plant(s). Various designers/licensors make offers on technology(s) which have been in operation for a year or two. Sometimes, however, as per logical analysis, the technology(s) appears to be sound and effective with a record already implemented but with a short operation experience—perhaps no more than a year or two. There is a trade-off

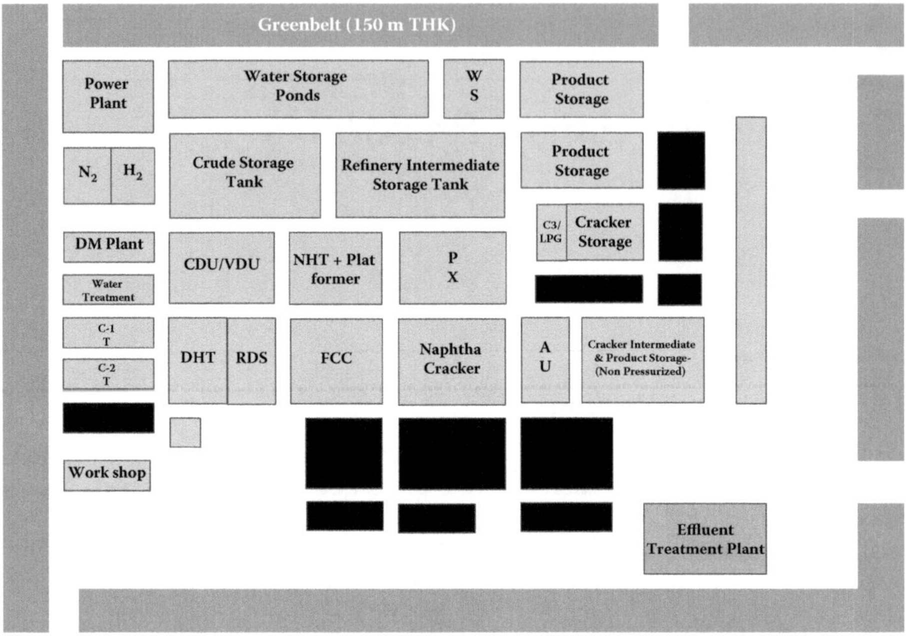

FIGURE 3.1
Legends: N_2: Nitrogen plant & storage; H_2: Hydrogen plant & storage; DM: de-mineralized water; CT: Cooling tower; WS: Water storage; C_3: Propane; LPG: Liquefied Petroleum Gas; CDU/VDU: Crude distillation unit/Vacuum distillation unit; NHT/Plat former: Naphtha hydrotreatment unit/Catalytic reformer unit; FCC: Fluid catalytic cracking; DHT: Diesel Hydrodesulphurization; RDS: Residue desulfurization; PX: Paraxylene; AU: Auxiliary units.

in this respect: if the technology selected is almost new (i.e., operational experience is less), there is a risk with respect to safety of which knowledge is less; but the user is enjoying the right to first access to the technology, thus enjoying competitive advantage over others in the business. In the industries presently, the requirement that the subject technology be in operation for 5 to 10 years, with a minimum of two years of reported successful operation is accepted by user industries.

2. It is prudent to go through pros and cons of the technology in operation through authorized agencies and through global network studies instead of depending on the information and records submitted by the technology supplier.
3. QRA (Quantitative Risk Assessment) and EIA (Environment Impact Assessment) should be carried out for the technologies through an authorized agency.
4. The technologies should fit in site selection like infrastructure, demography, skilled labor force, climate, seismic nature of the site, etc.

The technology selection process should not be restricted to process configuration only, but also be extended to selecting the offsite and utility technologies, such as an effluent treatment plant (ETP), flare, water treatment plant, dispatch facilities and internal power generation, if any.

3.2 Hazard Identification and Risk Assessment (HIRA)

As a preventive measure of minimizing the chances of accidents and reducing the possibility of injuries, loss of material and degradation of the environment, it is necessary to identify hazards and analyze, assess and control risks. Safety of the process plants can be achieved by auditing the plant design and operating the plant by following hazard identification and risk analysis techniques and adopting measures suggested by the analysis.

In India, the occupational Safety, Health and Chemical Hazards Sectional Committee, under the chairmanship of director general of National Safety Council formulated the code of practice on hazard identification and risk analysis. The chemical division council approved the same and the Bureau of Indian standards adopted it as IS: 15656:2006.

ANSI safety standards including OHSAS on occupational safety, health and chemical hazards being followed internationally are also available for reference and the Indian standard as mentioned above is in conformity to it.

The methodology in clause 1 of Indian standard helps the organization management in systematic identification of hazards and quantification of the risks associated with the operation of process plants. As per standard, certain industries are required to carry out risk analysis as a part of the statutory requirement and emergency planning.

Risk analysis / assessment is a process comprising of following four steps:

- Hazard identification
- Consequence assessment
- Accident frequency assessment
- Risk estimation

The standard describes the essential nature of each of the above steps and a variety of techniques for identifying hazards and quantification of accident consequences and frequency towards the final risk estimation. In clause 2, various technical terms used are defined. In clause 3, the methodology for carrying out risk analysis is given in the form of a chart as given in the following table.

Sl.No.	Project stage	Hazard identification/Hazard analysis techniques.
i.	Pre-design	a) Hazard indices, b) Preliminary hazard analysis, c) What if analysis, d) Checklists.
ii.	Design/ Modification	a) Process design checks and use of checklists, b) HAZOP studies, c) Failure modes & effects analysis, d) What if analysis, e) Fault tree analysis, f) Event tree analysis.
iii.	Construction	a) Checklist, b) What if analysis.
iv.	Commissioning	a) Checklists, b) Plant safety audits, c) What if analysis.
v.	Operation and Maintenance	a) Plant safety audits, b) What if analysis, c) Checklists.
vi.	Startup & Shutdown	a) Checklists, b) What if analysis.

Though the subject 'emergency handling' is not mentioned above, the same should also be considered while following sl. no. vi above.

In clause 4, regarding the life span of a process plant comprising a number of stages with respective hazards, hazard identification and risk analysis techniques are explained that should be applied at each stage, as enumerated in the table above.

Hence, to discuss hazard identification and risk assessment (HIRA), it is prudent to subdivide the subject into three categories: Hazard identifications and techniques (HAZID), Hazard probability/frequency of occurrence analysis of the hazard, and consequence/impact assessment—commonly called QRA (Quantitative Risk Assessment)—followed by risk estimation. These are discussed in detail as below.

3.2.1 Hazard Identification and Techniques (HAZID)

In the industries, there are hazards at three stages; namely, design and engineering of the installation, use of chemicals/hydrocarbons and the overall management system to manage all the risk involved in sustaining the operation and business.

Regarding design aspects, various references, standards and statutory obligations required to set up the industry are discussed in separate subchapters 3.4, 3.5 & 3.6. In this section, hazardous aspects of chemicals/hydrocarbons are discussed with respect to their harmful properties and their identification techniques and ranking. Following are the hazards of the chemicals and hydrocarbons used in the process:

- Flammability
- Explosiveness
- Corrosiveness
- Oxidizing
- Radio activity

- Heat sensitivity
- Water sensitivity
- Toxicity

NFPA (National Fire protection Association) provides the ranking of hazard in a scale of 0 to 4 with increasing intensity for three main hazards, namely health, flammability and reactivity. In general, major hazard potentials in petroleum/petrochemical industries are classified in two broad categories as follows:

- Fire and explosion hazards
- Toxic hazards

Identification of areas of vulnerability with respect to the above is most important for safety and loss prevention. Early hazard identification allows early mitigation measures with optimum impact on cost and time.

The objective of hazard identification is to identify hazards and unidentified events which would cause accidents. The hazards inherent to the process or in the plant are identified first and then the focus turns to the evaluation of events associated with the hazards. The hazard identification studies generate a list of failure cases by considering the following:

- Form in which chemicals are stored in the process
- Nature of hazard
- Quantity of material contained

There are two categories of hazard identification methods: namely comparative methods and fundamental methods, which are explained in the code.

Hazard identification is both qualitative and quantitative and highlights the risk of each single failure to focus separately on the subject, although in the actual process there may be multiple failures. The following are the two most important identification techniques and are discussed subsequently:

- Hazard indices and categorization
- Hazard checklisting

3.2.1.1 Hazard Indices and Categorization

It was developed by Dow Chemical Company to evaluate the relative loss potential of the process installation thus identifying and quantisizing the hazard in terms of ranking it. In earlier days insurances companies used to follow these techniques to identify and quantify the hazards. Following are different indices used in the facilities; finally, based on the values of the indices, hazard categorization can be identified using a table which also follows.

A. Hazard indices
   a. Dow index
   b. Toxicity index
   c. Mond index

a. **DOW index**

In this process, hazards are classified into three categories:

- General process hazard or base process hazards (BPH)
- Special process hazards (SPH)
- Material factor (MF)

Base hazards are identified in terms of whether the process involves exothermic or endothermic reaction(s), whether the process is carried out in an enclosed building or in open area, and whether the process involves handling or transportation.

Special process hazards include temperature, pressure, corrosion, erosion, leakage, flammability range and so on.

Material factors include quantity of hazardous material or its flow rate, flammability, reactivity and toxicity.

In the DOW index calculation, base process hazards (BPH) and special process hazards (SPH) are selected; then the values are obtained for the respective hazard known as hazard penalty factor, denoted by F_1 and F_2 for BPH and SPH respectively. There are normally assigned penalty factors for the operations with these hazards. The penalty factor(s), however, is to be modified depending upon what loss control parameter—such as emergency shutdown, cooling water system, and so on—are provided in the process. The penalty factor(s), generally vary from 0 to 1 with one scale, and in some scales devised by different institutions the figure may be more than 1. Here, 0 means most safe and the degree of unsafeness rises with the increase in the value of this factor.

The BPH penalty factor (F1) for combustion process is given by:

$$\text{Log}Y = 0.305 \times \text{Log}(H \times Q) - 2.965 \quad (1)$$

where Y = BPH penalty, Q = quantity of inflammable material in Kg-mole, H = heat of combustion in kJ/Kg-mole.

TABLE 3.1

Penalty factors for base process hazards (BPH)

Base Process Hazards	Penalty factor
i. Combustion	0.2
ii. Hydrogenation, hydrolysis, alkylation Isomerization, sulfonation, neutralization, i.e., exothermic or thermo neutral reactions	0.3
iii. Oxidation, Polymerization, Condensation	0.5
iv. Halogenations	1,0
v. Calcinations, Pyrolysis, Thermal cracking, Electrolysis	0.2
vi. Nitration	2.0

The penalty factor (F_1) for combustion process = 0.2 as seen from the Table 3.1. If special process hazard (SPH) is used, F2 should be taken from Table 3.2 on the following page. Similarly, material factor (MF) should be determined from Table 3.3.

From Table 3.3, for flammability (N_F) value 2 & reactivity (N_R) value 0, the MF comes out to be 10; for flammability value 3 & reactivity value 4, the MF becomes 40, and so on; i.e., MF can be maximum 40 and minimum 0.

Hence, the value of DOW from the above data can be calculated as follows:

$$\text{DOWindex} = (1 + F_1) \times (1 + F_2) \times \text{MF} \tag{2}$$

DOW index is commonly called as FEI; i.e., fire & explosion index. Its values above 95 fall under the severe hazard category, while values below 65 are in the very low risk category. However, risk category is not classified based on the DOW index only; the toxicity index also comes into play in hazard classification.

Example calculation of DOW/FEI:

In the hydrocracker unit, there is a possibility of exchanger rupture, causing a leak of hydrogen and hydrocarbon. Consequently combustion of hydrogen with atmospheric oxygen will occur instantaneously followed by combustion of hydrocarbon present along with hydrogen in that exchanger; there would be some ammonia also in that exchanger and hence, oxidation of ammonia would occur as well. Let us calculate DOW index, i.e., FEI in this case as below:

F_1 from BPH

The penalty factor for combustion from Table 3.1:0.2	
The penalty factor for oxidation from Table 3.1	:0.5
Total	:0.7

TABLE 3.2

Penalty factor for special process hazards

Temperature of handling	Penalty factor
Above flash point	0.25
Above atmospheric boiling point	0.6
At auto ignition temperature	0.75
At 0 to 30 °C	0.3
Below 30 °C	0.5
• At high pressure where P is relief valve set pressure in bar absolute & K is constant where K = 0.7 for highly viscous material = 1.2 for compressed gas = 1.3 for liquefied flammable gas = 1.0 for compressed gas released to atmosphere.	0.35 x log(P.K)
• At low pressure	
Hydrogen gas receiving system	0.5
Leakage	0.5
Vacuum distillation at <0.67 Bar	0.75
• Open system near to flammability	
Gas/air mixture	0.5
Process close to flammable range	0.75
Process in flammable range	1.0
• Corrosion / erosion	
<0.5 mm/yr	0.1
>0.5 mm/yr	0.2
>1.0 mm/yr	0.5
• Leakage	
Pump mech seal/gland packing	0.1
Flange leaks	0.2
Abrasive slurry	0.4
Sight glass/bellows/expansion joint	0.1
Welding joint	0.1–1.5

TABLE 3.3

Relation among material factor (MF), N_R which denotes reactivity and N_F which denotes flammability

N_R N_F	0	1	2	3	4
0	0	14	24	29	40
1	4	14	24	29	40
2	10	14	24	29	40
3	16	16	24	29	40
4	21	21	24	29	40

Note: Flammability depends on heat of combustion and reaction on decomposition temperature.

F_2 from SPH:

The process is a very high pressure process like at about 150 bar and has a temperature of about 150°C. Due to reverse joule Thompson effect due to sudden expansion of hydrogen leaking from high pressure to atmosphere results in auto-ignition of hydrogen regardless of temperature. With this information:

The penalty factor for auto-ignition from Table 3.2	:0.75
(For P=10 and K=1)	
The penalty factor for high pressure from Table 3.2	:0.35
Total	:1.10

MF value:

Hydrogen in the heat exchanger is in the molecular stage and not in the atomic stage. Hence, there would be no reactivity of hydrogen, but its flammability range is 4 to 75%, minimum value of 4% should be considered for making explosions in the presence of atmospheric oxygen.

Hence, in Table 3.3, for N_F value of 4 and N_R value of 0, the MF comes out to be 21.

DOW index/FEI:

Let us put all the values in equation (2) above as follows:

$$\text{DOWindex} = (1 + 0.7) \times (1 + 1.1) \times 21$$
$$= 1.7 \times 2.1 \times 21$$
$$= 74.97$$

Hence, DOW index/FEI is 75 for this scenario; i.e, it appears to be a serious hazard but not very serious as per previously mentioned definition on hazard severity. However, this evaluation resulted without considering other hazards in play. That is why the DOW index/FEI is an indicative hazard identification technique. To be more safe in evaluation, consequence modeling should be carried out in this case.

b. **Toxicity index (TI)**

It is obtained multiplying the risk of internal explosion by hygienic risk factor. As per NFPA, toxicity index is given by

$$\text{TI} = (T_h + T_s(1 + \text{BPH} + \text{SPH}))/100 \quad (3)$$

where T_h is the toxicity number derived from NFPA health factor, N_h or NFPA index, and T_s is the penalty factor derived from threshold limit value (TLV), a factor with respect to fatality.

TABLE 3.4

NFPA index co-relation

NFPA index	T_h	MAC, ppm	T_s
1	50	< = 5	125
2	125	5 – 50	75
3	250	>50	50

There is another parameter like MAC in ppm (parts per million) which means minimum anesthetic level in the toxic release. The table of correlation among all these parameter is shown in Table 3.4.

c. **Mond index**

It involves ranking of all internal hazards in the total process. For example, if an installation comprises various process units to complete its production process, then hazard ranking of all process units is to be carried out first. Then various alternative routes of production, say, by changing the configuration (i.e., by altering the process units), are to be compared with respect to hazard ranking in combination of all process units in the sets; then, overall ranking of hazard for the installation is to be done. This is generally carried out by bar chart preparation of hazard ranking for each process units in isolation and in combination within alternative sets.

B. **Hazard categorization**

Hazard is categorized based on DOW index and Toxicity index, as shown in Table 3.5.

3.2.1.2 Hazard Checklisting

Though checklist preparation is a qualitative approach in hazard identification techniques, it is a very important exercise in a very complex process and very much needed in the petroleum industries. As an initial activity, all documents such as plant layout/plot plan, process flow diagram (PFD), process piping and instrumentation diagram (P&ID), equipment data

TABLE 3.5

Hazard categorization

FEI (DOW index)	TI (Toxicity index)	Hazard category
< 65	< 6	I
95 = > 65	10 < = > 6	II
> = 95	> 10	III

sheets, engineering drawings and documents, and hazard area classification drawings are to be identified and made available on board. Now, check-listing is to be done as follows:

a. **PFDs review**

- Process flow scheme is well understood and it serves the intent of the process.
- All operating conditions like flow, temperature, pressure, heat duty and residence time (wherever required) are provided. These figures may be provided inside the diagram(s) or can be provided in the form of a table below the diagram or as an attachment to the diagram. If done so, all stream numbers are to be provided in the diagram(s) so that material balance and heat balance can be well understood.
- Legends should be provided for different operating parameters.
- All process controls are provided that serve the purpose with respect to both process operation steadiness and safety.
- In addition to control instruments, all monitoring instruments, at least for heat balance, are provided wherever required. Additional monitoring instruments may be checked during P&ID review.
- All safety valves are shown along with set pressures to check whether these are reasonable and below design pressure, as per norms.
- Safety valves which need to stand by or along with pilot valve, with or without isolation facility, with rupture disc for viscous/dirty fluid service should be indicated. Please note, when the main safety valve is provided without standby but with pilot valve, there should be an isolation valve for the pilot valve and no isolation valve for the main safety valve; however, where standby safety valves are provided, isolation valve can be provided for both safety valves.
- Flare connections to the equipment should be indicated but it should be reviewed in detail during P&ID review.
- Safety design criteria, i.e., whether it is a case of cooling water failure or a case of fire should be indicated in PFDs notes or as a basis attachment.
- Check valves / non return valves (NRV) have been provided to avoid the situation where reverse flow can cause hazard. But check valves for all rotary discharge end can be reviewed during P&ID review.
- Vents, drains are provided wherever required. However, these are not foolproof in PFDs; the same is to be ensured during P&ID review.
- All equipments isolation and/or bypass facilities are provided to facilitate handing over the equipment to maintenance as well as

to handle unsafe situation if it is found necessary. However, all isolation, bypass are subjects of P&ID review.

- Grade level of the equipment should be provided. Sometimes, it is mentioned in PFDs, but it must be checked during P&ID review.
- Utility connections for start-up and shutdown requirements, as well as for catalyst regeneration processes should be provided.
- Separate flow diagram(s) for catalyst regeneration process if required is/are provided, along with control and monitoring instruments for heat balance.
- Heat balance and material balance charts or data sheets should be provided as an annexure of PFDs.
- Slop routing facilities should be indicated.
- Heat tracing and insulation philosophies should be mentioned as a note in PFDs, which can be checked in P&ID review.
- Quality control analytical procedure with frequency of testing, startup and shutdown philosophies and catalyst regeneration details should be provided as a detailed document along with PFDs.

b. **P&ID review**

- Line sizes and line identity numbers, along with metallurgy classification codes are to be mentioned in each line; heat tracing and insulation (if required) should be mentioned in line identification, as mentioned above; a table for decoding metallurgy should be provided.
- Standby equipment should be reflected in the drawings.
- All HPV (high point vent) and LPD (low point drain) should be indicated, as per process requirement.
- Double isolation with in-between blind and bleed facilities for all piping should be indicated in the drawings wherever needed; namely in the battery limit and also inside units such as separation of low pressure zones from high pressure zone, and, in some special cases, as decided on the basis of design to allow positive isolation during shutdown.
- Specially required line slopes should be indicated wherever applicable.
- Safety valves with standby will have isolation valves at upstream and downstream of the safety valve; similarly, safety valve with pilot valve will have isolation valves at upstream and downstream of the pilot valve; also, safety valve with rupture disc at upstream of safety valve used for viscous or dirty service will have isolation valve at upstream of rupture disc. One pressure transmitter should be provided in between rupture disc and safety valve to identify whether rupture disc has ruptured or not.

- For boiler steam drums, numbers of safety valves are provided with a pop setting at different pressures; there should be no isolation valve at upstream of the highest pressure rating safety valve; however, boiler inspection systems of respective countries don't allow any isolation valves, even in a series of popping safety valves at low pressure.
- In offsite P&IDs, as the lines are long, thermo safety valve (TSV) with isolation valve lock open with outlet connected to storage tank should be provided.
- Control instruments for temperature like TC should be provided, along with a legend for field control or remote control from the control room.
- Similarly, control instruments for flow and pressure control like FC and PC respectively should be indicated along with a legend for field control or for remote control from the control room. For pressure control, in some exceptional cases, there is no scope to change control set point by the operator except by the instrument engineer with approval from a competent authority; such instruments called PCV (pressure control valve) should be indicated with a proper legend.
- In some cases, shutdown valve (on/off type) may be required as per severity of the process, for example, for flow from a very high pressure to low pressure zone. The same should be indicated in P&IDs.
- Measuring / monitoring instruments for temperature like TG, RTD, TW, etc., for flow like DP, US, VM, and mass flow meter, for pressure like PT, DPT(with or without diaphragm), PG, and for level like float type, DP, LT (magnetic, Ultrasonic, radio isotopic), etc. should be indicated.
- Control loops close or open type should be properly indicated with special indications of single loop control or split control or ratio control or other logic control like max., min., and so on.
- Air to open or air to close functioning of control valves should be indicated to understand the safety in operation.
- Heat tracing / insulation requirement for instrument impulse lines should be indicated.
- Leak class of control valves like class 3, 4, 5, 6 are to be mentioned; for example, in a control valve placed in a liquid line where flow occurs from a high pressure to low pressure circuit there may be a chance of high pressure gas coming out to a low pressure circuit in abnormal situations such as a very low level in upstream equipment; a control valve with class 3, 4, 5 would endanger the downstream with an escape of gas through the control valve; class 3 is most dangerous and so on; in that case, a

class 6 control valve would be selected. These are all covered in SIL (safety instrumentation level) systems. Hence, instruments in P&IDs should be as per selected SIL norms.

- To cover special safety precautions as per selected process, there may be a provision required to supply the pneumatic / hydraulic pressure to the shutdown valve when there would be air supply cut off to the valve in sudden power failure or any other emergency shutdown. Also, as per process logic, the shutdown valve should be either fully open or fully closed. To ensure this operation in such case, a pneumatic or hydraulic cylinder of desired capacity and pressure with adequate thermal insulation against damage from fire should be provided close to the shutdown valve; the same should be indicated in P&IDs.
- Size of control valves, shutdown valves along with metallurgy and class specification should be indicated in P&IDs.
- Metallurgy and other specifications of the measuring instruments should be provided as separate documents ,along with P&IDs.
- Wherever, control valve bypass facility is required should be indicated in P&IDs.
- All logic control diagrams like voting, any algorithm like sum, >, <, min, max, etc., along with trip logic should be provided along with P&IDs and notes should be provided in the respective P&IDs with proper symbols against such control loops.
- ROVs, MOVs are generally required in offsite storage facilities; those should be indicated in respective P&IDs.
- All alarm facilities, along with types of switch selected should be indicated in notes of P&IDs so that these can be taken care of in the DCS console system.
- Hand jack facility to control valve to operate the same manually should be indicated wherever it is required or not, as per process design basis.

c. **Engineering review**

- Project design standards / basis should be available.
- Mechanical engineering:
  - In line with P&IDs, all line numbers including size, class, with or without heat tracing and with or without insulation (hot or cryogenic) should be provided. As such, engineering drawings should match with PFDs and P&IDs, as discussed above.
  - All reducers, expanders, elbows, flanges, blinds (tail or spectacle type), high point vent (HPV), low point drain (LPD) not only to be provided as per P&IDs, but also to be provided as per construction and maintenance requirement and for construction and maintenance needs as decided during engineering.

- Gaskets, whether metallic (spiral wound, tongue and groove) or non-metallic (asbestos, permanite) and stud bolts standards should be reflected in engineering documents.
- Requirement of earthling continuity strips across flanges connected by non-metallic gaskets for long lines going up to storage tanks should be reflected in engineering documents.
- Stud-bolts standards, specifications should be indicated.
- Standby equipment should be reflected in the drawings in line with P&IDs.
- Double isolation facilities with the blind and bleed system, bypass valves facilities, check valves with direction indications, all flanges wherever required, check valves wherever specially required as per PFDs and P&IDs, as well as all check valves at discharge of rotary equipment should be indicated.
- Isolation valves for safety valves should be placed in such ways that the valve wheel is located downward; the same logic also should be applicable in any vapor isolation valve and cooling water isolation valve because external air-water ingress to valve glands in case of valves filled with vapor and not with liquid, and air ingress in case of cooling water valves may cause corrosion and may result in detachment of the valve spindle during closing of the valve. This would make them unable to be opened further, with the result that the whole unit needs to be shutdown. This aspect should be indicated in the drawings.
- Grade levels of all equipment should be indicated.
- 3D isometric drawing should be provided to understand operability level of suction/discharge valves of rotary equipment, along with plan/elevation of all valves at all levels/platforms should be indicated to understand operability of the valves and blinds, as well as to demonstrate minimum distances are ensured wherever required.
- Welding joints standards along with welder qualifications should be included in the engineering in the quality assurance section, where all other aspects like hydrotest, radiography and the like are covered.
- Slopes required in some lines should be visible in 3D isometric drawings.
- Spec. breaks should be shown properly in the drawings.
- Mandatory spares inventory should be reflected in the documents.
- General layout, i.e., GA drawings should be available to understand the layout of all lines, including A/G (aboveground) and U/G (underground) lines.
- GA drawing should be checked for process units and offsite against all utility lines such as cooling water, fire water, flare piping, closed blow down (CBD), oily water swage (OWS)

leading to ETP, process water treatment, DM water, drinking water, boiler water, plant air, instrument air, inert gas, and so on.

- Fire water network should cover all roads, tank farms, ETP, Power plants, electric sub stations, Flare areas, DM water plants, dispatch facilities, process cooling towers and process units with fire water hydrants and hose boxes at regular intervals, as per norm. Fire water risers should be provided at elevated areas in process units.
- Similarly, network drawings for all utilities in the offsite should be provided.
- Proper wrapping/coating should be provided for the pipes laid underground.
- Numbering of all drawings, equipments should be indicated, along with legends.
- Vendor scopes are properly identified wherever needed, in respect of package items.
- CBD (closed blow down), open drain to OWS (oily sewage system) should be indicated, as per requirement.
- provision of flare knock drum and CBD vessel inside the process units should be indicated in the drawings.
- All structural drawings should be in conformity with piping GA drawings to avoid mismatch in piping layout in pipe racks. Lines' tier system should be well demonstrated in the drawings to understand the efficacy of the engineering with respect to maintenance and inspection.
- Thermo-wells reinforcement pads wherever required should be reflected in the drawings.
- Shoes for piping in pipe racks should be indicated in engineering.
- Vertical pipe support springs and horizontal pipe spring supports wherever required should be indicated, along with lock systems which should be unlocked before hydro test and commissioning.
- For long length hot utility lines at offsite, engineering care should be taken to provide expansion loops. For both hot and cold utility lines, say, steam, cooling water, service water, plant air, instrument air, boiler water, fire water lines and so on, adequate numbers of HPV & LPD should be provided, along with isolation valves. The lines should be well supported whether at grade level or elevated.
- Fire water network should cover all peripheral roads surrounding process plants, offsite storage areas and design of the network, along with fire water pumping systems and should cover firefighting emergencies like operation of pumps and maintaining header pressure as per statutory obligations.

    - Flare piping, if provided with a single header, should cover the discharge need both from high pressure (HP) safety release and low pressure (LP) safety release from process units. The header, well supported with LPDs at regular intervals of distance, should be included with pilot flare header, and the local panel should be provided at flare stack location where flare knock out drum would be there with auto pumping from the knock out drum to ETP during any emergency of liquid carry over to flare header. Flare stack should be provided with a hydraulic seal drum at its bottom. A flare stack local panel should be provided with a flare light up system.
    - Separate control should be provided for ETP and water treatment plants.
    - Minimum distance criteria between control valve and orifice flange, control valve to bend, quill to bend, and so forth should be reflected in engineering drawings.

- Civil engineering:
    - Structural stability calculation duly approved by competent authority should include in engineering.
    - Structural specifications, drawing; namely a plan tier wise and with elevation should be available.
    - For pipe rack systems and technological platform (platform holding series of equipment at an elevated position) construction depending upon load studies, RCC structure should be done if required; if any steel structure is there at grade level, fireproofing is to be done as per statutory norms.
    - For offsite piping support, RCC type should be done as per load of the piping system; for one or two piping with minor load, PCC support can be done as per load calculation, which is to be checked.
    - Operation control room construction should be blast-proof type as per hazard and risk analysis of the surrounding process area outside. However, it may be mandatory in certain cases as per statutory norms; hence, norms are to be checked first.
    - For foundation construction of equipment, care should be taken with respect to seismic zone;, i.e., for that area piling and raft foundations would be required, as per engineering and statutory guidelines.
    - For storage tank construction, foundation criteria against seismic zone and cyclonic zone is to be checked; piling test by authorized agency and subsequent piling, if required, should be carried out as per procedure.
    - Requirement of monkey ladder, platform, and stair cases should be identified.

- Equipment insulation should be properly indicated /mentioned in the drawings.
- For some coolers, insulations are not provided to enhance efficiency of cooling; in such cases, coolers should be properly painted and be caged with net type structures around the shell of the coolers to avoid human burn injury.
- All piping, equipment, structural support painting should be as per project standards, which should be demonstrated through documents.
- Underground trench drawings for cable laying should be demarcated for electrical and instrument cables. Document should be reflected in sand covering of the trenches to avoid water accumulation in the trenches.
- OWS, units surrounding surface drains and storm water drain network drawings should be available, along with oil catcher facilities in storm water drains.
- The pipes crossing roads through culverts should be reflected in engineering.

- Instrument engineering:
    - All instrument locations should be indicated properly in line with PFDs and PIDs; the operability and maintainability of the instruments should be demonstrated in the in engineering documents.
    - All control valves, measuring instruments and data sheets should be provided.
    - All interlocks logic diagrams and hardware engineering should be demonstrated.
    - Orifice tapings for liquid services should be inverted 'V' shaped; while for gas services, it should be installed in real 'V' shape direction due to respective properties of liquid and vapor; this should be reflected in the drawings.
    - Engineering documents should reflect use of all instrument fittings like junction box, barrier, converter, glands, seal, and so on.
    - Instrument impulse lines length should be as minimum as possible; also, these lines should be heat traced and insulated wherever required. This should be reflected in the drawings.
    - Special care should be taken in engineering of small tapings which sometimes are ignored. Pressure and flow impulse taping are vital, and standards should be reflected in engineering; sometimes, the joints found give away causing hazards.
    - Thermo well insert direction inside the pipe should be at 45 degrees angle in at least liquid service to avoid thrust on the thermo couple.

    - Thermo well and thermo couple metallurgy should be as per engineering standard.
    - Safety PLCs (programmable logic controller), along with redundancy should be provided.
    - Dedicated instrumentation for burner management systems should be provided.
    - Engineering supervisory control rooms in addition to the operator control rooms, should be included in the engineering.
    - Instrument cable duct engineering should be as per standard to ensure proper identification and replacement / repair.
    - Instruments cabinet rooms and their accessories layout should be demonstrated.
    - Battery limits should be properly defined.
    - FAT (factory assessment test) and SAT (site assessment test)- documents should be available for DCS.
    - Equipment running / stop indications should be available in DCS console engineering.
    - Engineering on all hardware alarms and trip devices in the operator console should be as per standard.
    - Inert gas depressing system to inertize control room in case of fire / smoke should be available in DCS control room, if required by statutory obligation.
    - Fire/smoke detector alarms should be provided in the control room.
    - Gas detectors should be provided in process units and storage tank farms as per hazard analysis by authorized agency.

- Electrical engineering:
    - Electrical hazardous area classification should be available.
    - Selection of induction and synchronize motors should be well demonstrated to maintain power frequency and power factors as per standard.
    - All motors data sheets should be provided.
    - All electrical substations, field equipment like transformers, lighting, etc., cabling (power cable and control cable), joints, switch, glands, layout, and installation documents should be reflected as per project standards.
    - Black start / emergency DG set should be provided to connect emergency process units and services.
    - All equipment, storage tanks and building earthling connections, along with earthling pits should be reflected in engineering.
    - Electrical relays and control circuit diagrams should be available.
    - Total lighting documents for equipment, all unit fields and offsite, all control rooms, substation streets, tank farms, roads, buildings and so forth should be available.

- Air conditioning / exhaust facilities documents should be provided for control rooms and for office areas; standby refrigeration equipment, along with details of auxiliary cooling systems like cooling towers or air fans should be provided.
- For existing facilities, electrical system audits should be carried out and reports should be available to monitor the compliances of recommendations.
- Requirement of public announcement (PA) system should be reflected in engineering drawing and document.

d. **Lay out plan / plot plan**

- Plot plans for the all the following should be available:
    - Overall plot plan for the industry battery limit (enclosure).
    - Each process unit plot plan.
    - Plot plan for fire and water networks, along peripheries of all roads, process units, offsite storage tanks, bullets/Horton sphere areas, ETP, flare area, water treatment plants, power plants (if there), substations, product dispatch areas, all office buildings and control rooms, along with fire water reservoirs and pumping facilities.
    - Plot plan for cooling towers and circulating water networks through the industry.
    - Plot plans should be available for all utilities generation plants like plant air, instrument air, DM water, water treatment plants and effluent treatment plant (ETP).
    - Plot plan for flare area.
    - Plot plan for internal power plant, if provided.

- Equipment inter distances in plot plans should be as per statutory norms. The distances should be indicated with drawings provided with scales of distances.
- Plan for elevated tier levels and underground levels should be reflected in the layout.
- Equipment in layout should match with PFDs and P&IDs.

e. **Equipment-wise checks**

- Vessels:
    - Data sheet for vessel should be available.
    - Type of vessel (vertical or horizontal), along with dimensions and its elevation from grade level should be indicated.
    - Type of vessel support should be indicated.
    - Vessel platform requirement, if any, should be indicated at the top.
    - Provision of nozzles for-inlet, type of entry (whether tangential or axial), outlet, drain, steam out, vent, relief, utility connections, manhole, overflow (if required), level tapings

separately for indication / trip, pressure measurement, temperature measurement, etc. should be available.
- Requirements of painting, heat tracing/jacketing, and insulation should be indicated.
- Inlet distributor, outlet collector (if any) should be identified.
- Internal partitions in the vessel, if any, should be indicated.
- Requirement of internal linings—stainless steel, cement, epoxy coating, guniting, etc., should be indicated.
- Demister pad, if required, its size and specification should be specified.
- Vortex breaker with particle arresting net, if any, should be specified.

- Reactor:
  - Data sheet for the reactor should be available.
  - Dimension of the reactor, along with elevation from grade level and insulation requirement along with its specification should be indicated.
  - Inlet nozzle, entry type, collector, distributor, internal lining should be specified.
  - Nos. of catalyst beds, each bed height and bed mechanical strength should be specified.
  - Quench nozzles, distributor, outlet nozzle, outlet collector / particle arresting net in the bottom nozzle should be specified.
  - All instrument connections should be indicated.
  - All utility, vent, drain connections should be indicated.
  - Platforms required at various levels should be indicated.
  - Relief valves with connections, isolation, and emergency depressurisation (if any) should be indicated. Generally in the process, reactors are connected through a series of exchangers, coolers without any isolation in the series to a vessel where pressure control facility is provided. Hence, relief valve emergency depressurization can be provided in that vessel to take care of the whole series in the upstream connected without isolation.
  - Liquid vapor disengagement trays wherever required should be indicated.
  - Support, platforms at various levels should be indicated.

- Columns:
  - Data sheet for the column should be available.
  - Dimensions along with elevation from grade level and insulation requirement along with its specification should be provided.
  - Provisions for nozzles; namely feed inlet, type of entry of inlet nozzle, product outlets, side draws, vapor returns, top

reflux inlets, pump around draw and returns, drain, steam out, vent, relief, manholes, bottom nozzle, level taping separately for indication / trip, temperature measurement, pressure measurement, and so on should be indicated.

- Requirement of painting and insulation, if any, should be indicated.
- All distributors for feed inlet, reflux inlet and pump around return, should be defined.
- Partition chamber in the bottom, whether open or sealed up to the bottom most tray, should be defined.
- Bottom reboiler type; namely vertical or horizontal, natural or thermo-syphonic, once through or circulating, for which there would be two or one nozzles respectively at the column bottom.
- Bottom outlet collector / vortex breaker should be specified.
- Type of internals; namely tray type, nos., number of passes, metallurgy, chimney trays, along with dimensions should be available. If packed beds are provided, packing type, dimension, material of construction, number of packed bed, random packing or structured packing, packed bed heights, collector and distributors before first bed and in between beds should be indicated.
- Requirement of demister pad, along with its location, dimension and material of construction (MOC) should be specified.
- Requirement of vortex breaker to be specified.
- Requirement of platforms, monkey ladder or stair case should be indicated.

- Heat exchangers / coolers / condensers:
    - Data sheet for the exchanger should be available.
    - Type of heat exchangers to be specified in line with PFDs and P&IDs.
    - Number of tubes, tube diameter, length, MOC, pitch (triangular or square), number of passes both in tube and shell side, number of shell side baffles with layout and pitch, shell dimensions, MOC, dished end dimension, type whether fixed or floating head, channel head dimensions with inlet, outlet and other nozzles (vent, drain, flushing), exchanger whether single or twin should be specified.
    - Isolation facility should be indicated with bypass facility (when needed).
    - Painting, insulation should be indicated.
    - Instruments connections like TI, PI, TG connections should be indicated.

- Air coolers:
  - Data sheet should be available.
  - Dimensions with plan, elevation, inlet / outlet nozzles, number of tubes, tube diameter, type of fins fitted with the tubes, MOC (material of construction) of tubes, fins, inlet / outlet channel boxes to be indicated.
  - All drains, vent and instrument taping details should be indicated.
  - Isolation facilities should be indicated.
  - Nature of draft, i.e., natural draft or forced draft, should be indicated; if forced draft, details of driving facility (such as motors) with respect to number, capacity and other specifications of the motor should be provided.
  - Control facility like temperature control whether it is on bypass or with variable speed motors or varying the numbers of motor running should be indicated, along with local control or remote control facility. In any case, there should be local switch buttons to start/stop the motors.
  - Sometimes, there is a temperature control facility provided, as in the Louvre, i.e., keeping the air-fin cooler inside a box which has movable guide that vanes at the inlet to control air ingress. Sometimes, air ingress control is done by fan blade angle varying.

- Pumps:
  - Pumps data sheets along with characteristic curve should be available.
  - Type of pump (centrifugal or reciprocating or screw or lobe or gear), number of stages, MOC, its required min. NPSH (net positive suction head), and desired NPSH, discharge head, fluid property, minimum turn down ratio, efficiency, its driving unit (motor or steam turbine) details with overall efficiency should be indicated.
  - Rated, minimum and maximum impeller diameter, MOC should be provided.
  - Mandatory spares on pump impeller, casing, mechanical seal should be indicated.
  - Pump suction / discharge nozzles, suction strainer, discharge check valve, isolation valves, drain, vent, pressure gauge taping, pump foundation details and elevation from grade level should be indicated.
  - Pump-motor bearing details, pump mechanical seal API class and details to be indicated; for water pump, gland packing is provided to reduce cost; in that case gland specification to be indicated.

    - Relief valve details, generally given for positive displacement pumps as mentioned above, should be indicated.
    - Relief valve details of turbo drive along type of drive, i.e., fully condensing or extraction type along with steam entry trip lever mechanism should be provided.
    - If turbo driven, generally two lube oil pumps are provided for a big turbine; in that case, one lube oil pump should be motor driven and the other should be turbo driven; the details of these auxiliary sets should be available.
    - Turbine bearing mechanisms, particularly for heavy duty turbines where journal bearing is provided, should have - an emergency trip mechanism in case of failure of bearings.
    - Turbine trip against overspeed should be indicated; this may be a mechanical trip device with liver system but additionally, an electronic trip system also should be there which in present days is found to be more reliable.
    - There should be an operating panel in the field for heavy duty pumps and turbines, and facilities in the control room for monitoring and emergency stops should be provided.
    - In a reciprocating pump, vibration protection facilities should be indicated.

- Compressors:
    - Compressor data sheets should be available.
    - Type of compressor (centrifugal or reciprocating or screw), number of stages, MOC, number of impellers, impeller diameter, MOC, labyrinth seal, suction pressure, discharge pressure with differential head, fluid property, minimum turndown ratio, efficiency, type of driving unit (motor or steam turbine); details with overall efficiency should be indicated.
    - Compressor suction / discharge nozzles, suction strainer, discharge check valve, isolation valves, drain, vent, pressure gauge taping, compressor foundation details and elevation from grade level should be indicated.
    - Details of suction, discharge loading / unloading valves like specification, MOC, spares provision, stuffing box details should be provided for reciprocating compressor.
    - All instrument taping should be indicated.
    - For a reciprocating compressor, venting facility from a stuffing box should be indicated, along with details like manual venting to atmosphere or through control using inert gas as purging medium under differential pressure control (DPC) between inert gas and leaking gas in stuffing box.

- The auxiliary equipment details of the compressor such as lube oil and seal pump, along with their drive details (one motor driven and other turbo driven) should be provided.
- Compressor trip system like trip at high discharge pressure, high discharge temperature, high lube oil / seal oil temperature, low lube oil / seal oil pressure, compressor suction / discharge knockout vessel high level should be available.
- If turbo driven, generally two lube oil pumps are provided for a big turbine; in that case, one lube oil pump should be motor driven and the other should be turbo driven; the details of these auxiliary set should be available.
- Turbine bearing mechanisms, particularly for heavy duty turbines where journal bearing is provided, emergency trip mechanism should be available on failure of bearings.
- Turbine trip against overspeed should be indicated which may be a mechanical trip devise with liver system but additionally an electronic trip system also should be there which in present days is found to be more reliable.
- Compressor motor bearing details along with trip at high vibration and pump mechanical seal class and details to be indicated; for water pumps, gland packing is provided to reduce cost; in that case, gland specification is to be indicated.
- Relief valve details, generally given for positive displacement pumps as mentioned above, should be indicated.
- In case of centrifugal compressor, the type of antisurge, along with its detailed specification should be indicated; additionally, if spill back facility is required, that should be spelt out.
- There should be an operating panel in the field for heavy duty compressors and turbines and facilities in the control room for monitoring load variation; other major control and emergency stops should be provided.
- Vibration protection mechanisms like trips with vibration-like axial or radial should be provided. Generally, at compressor end, thrust bearings are provided and in turbine end journal bearings are provided. Tripping on vibration is required to protect bearings and the machine.
- Hazardous area classification for the drive should be specified.

- Fans / Blowers:
  - Data sheets should be available.
  - Dimensions along with all vane details and nozzle sizes and insulation requirement along with its specification should be available.
  - Details on suction, discharge damper for isolation. Suction strainer to be specified.

    - Drive mechanism whether motor driven or turbo driven should be available with detailed specifications.
    - Flow control facility like varying motor speed or governor oil pressure control in case of turbo drive should be specified.
    - All instruments taping should be indicated.
    - Type of turbine (API-611 or 612) selected should be specified along with details on auto start/cold start/hot standby modes.
    - Foundation details should be available.

- Fired heaters:
    - Data sheets should be available.
    - Heater duty, efficiency, fluid mass and heat balance should be reflected in the data sheets.
    - All zones; namely convection, radiation, super heater, economizer should be indicated in drawings and the data sheets along with insulation specification.
    - Tube diameter, length zone wise, number of tubes, type of furnace (namely vertical, cylindrical or box type) should be indicated.
    - Explosion doors should be available.
    - Fuel firing details like fuel oil, fuel gas, number of burners, burner nozzles dimension and details, purging and atomizing steam provision should be provided.
    - Furnace tubes steam purging facility in case of tube failure should be provided.
    - Steam/air decoking facility should be provided along with instrumentation.
    - Temperature control scheme should be provided, along with details on control equipment.
    - Skin, box and stack temperature indication in multi locations should be available.
    - Stack damper opening/closing damper position physical, as well as instrument indication to control room should be available.
    - Furnace draft indications should be provided.
    - Safety interlocks for emergency shutdown such as fire cut off, then feed cut off should be provided.
    - In case the furnace is provided with a balanced draft system,—i.e., with both forced draft air fan (FD fan) and induced draft fan for flue gas (ID fan)—with air preheater (APH), the stack damper auto opening facility should be included in furnace emergency shutdown, as mentioned above. In that case, in furnace safety interlocks as mentioned above, fire cut off with combustion air failure(i.e., FD fan tripping) should be provided with stack damper opening immediately.
    - In balanced draft, two FD fan should be provided, with both running at 50% capacity' and with one tripping, the other's

capacity should increase to 100% in autos; that should be provided along with all fans details.
    - Heater insulation and inside refractory bricking details to be indicated.

- Storage tanks/bullets/Horton spheres:
    - All respective data sheets should be available, including all dimensions including type of tank like fixed roof or floating roof.
    - Layout of all tanks should be as per statutory obligation.
    - Class A & B hydrocarbons are kept in separate tank farms. Class C and others categories can be kept together in same tank farm but not to be mixed with A or B categories.
    - All classified tanks are to be enclosed in a dyke wall made of earth or concrete as per safety calculations and standards for holding the content of the largest tank spillage.
    - Tank elevation from grade level should be available.
    - Tank bottom pad details should be available.
    - If steam coils are there in the bottom, all details should be available.
    - All structural details like roof support for fixed roof tanks, and for a floating roof all engineering details including pontoons, tank shell course number details, metallurgy, stair case details, flexible ladder details on floating roof, roof drains for floating roof with details like joints (swivel joint or pivot joint), floating roof seal, tank manholes (manhole numbers depend on the size of the tank), all inlet/outlet/circulation/drain nozzles, tank dip hatch, vent with flame arrestor for fixed roof tank, all level and temperature indication facilities, lightening protection, painting, insulation to be indicated wherever required.
    - Anti corrosion painting inside tank up to a minimum height to be provided to take care of situation like water ingress in the tank; for hydrocarbon like aviation turbine fuel (ATF), it is mandatory to carry out full anti corrosion painting inside.
    - Detailed drawings/engineering documents should be available for pumping facilities from and to the tanks.
    - For storing pressurised liquid in bullets or Horton sphere, no enclosure like dyke wall is necessary as contained hydrocarbon if leaks would not accumulate but however, it poses other threat of accumulation at ground at a further distance if the leaky vapor is heavier than air, for e.g., LPG, propane, propylene which is stored inside the petroleum industries. Here, installation of gas detector in all possible location of spread, as per hazard study like using dispersion modeling, should be done; engineering should demonstrate the same.

- Detailed engineering drawings/document should be available for pumping facilities from and to spheres and bullets.
- the safety relief valves of the bullets and Horton spheres should be multiple in number without isolation valve and feasibility of connecting the flare outlet to common flare instead of cold venting should be demonstrated.
- Horton sphere hot insulation to be reflected in engineering to take care of sun's heat.
- ROVs in sphere, bullets inlet, outlet, circulation lines (if any) should be in place as per statutory norms.
- All level, temperature indications should be reflected in engineering as per statutory norms.
- LPG bottling, if installed should have a detailed engineering document; it should be provided with auto filling, cylinder leak detection facilities, sick cylinder evacuation facility, with roller conveyer facilities up to loading truck, and so on.
- LPG bottling area floor should be masticated to avoid frictional spark during manual cylinder movement.
- Engineering should reflect good illumination in tank farm and spheres/bullets storage areas.

3.2.2 Hazard Probability/Frequency of Occurrence Analysis

There are two commonly used basic tools; namely Fault tree analysis and Event tree analysis, used for carrying out hazard probability analysis either to find out probability of a root cause to be responsible for an accident or probability of consequences arising from an unsafe act. These are discussed as follows.

3.2.2.1 Fault Tree Analysis (FTA)

Fault tree analysis (FTA) is basically a tool for root cause analysis of an event; i.e., it starts with the effect and is analyzed upward to find the root cause, the initial incident. It is a top-down logical diagram explainable through Boolean algebra to determine the relationship between critical events and their causes. The causes also can be termed as basic events. In another way, it can be said that FTA identifies, develop models and evaluate the unique inter-relationship of events leading to:

- Failure
- Undesired events/states
- Unintended events/states.

The methodology is defined, rigorously structured and easy to learn, perform and follow. Along with Boolean algebra, it utilizes probability theory, reliability theory and logic. It however follows the laws of physics, chemistry and engineering as well. It is explained in Flow chart 3.1 through a simple electrical circuit of a ceiling fan with one switch in the board and with an electric supply line for power connection as follows:

There are two methods of analysis; one way is checking all the possible causes of failure with physical testing wherever possible, event records and scientific reasoning.

By carrying out a megger test for electrical circuit, it was established that the continuity of wire was all right. The switch was checked and found to be all right. Power supply failure did not occur. Hence, the cause could be either due to the fan bearing ceasing or condenser failure. Generally bearing cease gives a pre-indication of noise during the running of the fan before failure; event records did not say so. So, the most probable cause of failure was the fan condenser stopped functioning. The suspicion was confirmed after checking the condenser.

Another method of analysis is with the use of probability functions; this method is followed when adequate event records are not there and adequate tools and physical testing resources are not there. The method can be explained as follows:

Event combinations that can cause the most undesired events to occur can be one subset alone or different subset combinations; accordingly, probabilities will vary due to multiplication of respective probabilities which may be available from the quality data of the cause contributing spares/parts/equipment with respect to probability of failure as per their quality control system and performance feedback system. In the present scenario, the probabilities are found as per industry resources as shown in Table 3.6.

Here, as a standalone, the most probable cause of the fan to stop running should be subset B; i.e., the fan condenser ceased. However, the probability of the combined failure of two subsets, say failure of both bearing and condenser, decreases and is also reflected by the probability which is obtained by multiplying the probabilities of subsets A & B.

In actual experience, the problem is a complex one with a number of basic events/subsets leading to the ultimate critical event. Now, let us carry out an FTA with such a complex system as explained below:

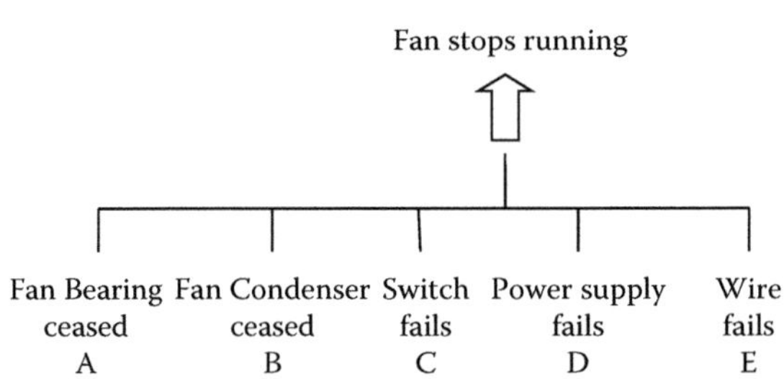

FLOW CHART 3.1
Each basic event (A, B, C, D, E) above is called sub-set (SS).

TABLE 3.6

Sub-set and its probability

SS	Probability
A	P(A)=1.0X10^-4
B	P(B)=1.0X10^-3
C	P©=1.0X10^-7
D	P(D)=1.0X10^-5
E	P(E)=1.0X10^-9

The crude distillation unit (CDU) of one petroleum refinery caught in a huge fire damaging most of the overhead piping network making the duration of shutdown for repairs about one month in order to recover from the damages. On survey after the fire was extinguished, it was found that the main crude oil feed pump was ruptured, and a total of about 25 products, intermediates liquids and gas lines got ruptured and were damaged in multi locations. So, here, there are two consequences: one is a crude pump and the other is a bunch of pipelines. To start FTA, let us start with the consequence; say, failure/damage of the crude pump, as explained through the Flow chart 3.2.

Conclusion of Flow chart 3.2: The root cause was the incorrect operation of the crude pumps or the selection of pump type was wrong. As such, two stage crude pumps should be selected for such high delta pressure differential between suction and discharge of the pump.

Excel based software is used presently for FTA which is customized for different industries and is available in the market.

The application of FTA can be used for:

1. Root cause analysis
2. Risk assessment
3. Design safety assessment

FTA is an analytical tool with advantages and disadvantages as follows:

a. **Advantages:**
    - It is basically a root cause analysis (RCA) where the objective is to find out basic event from the known consequence(s).
    - It not only helps in the investigation of the root cause of accidents but also helps in making decisions for design/procurement by applying RCA on the failure of the proposed equipment with various alternative criteria selection.
    - It is methodical, structured, graphical and quantitative and is used to model complex systems. It also covers hardware, software, humans, procedures, timing, and so on. The user only needs to know how, why and when the tool should be used. It is also used for systems evaluations such as:

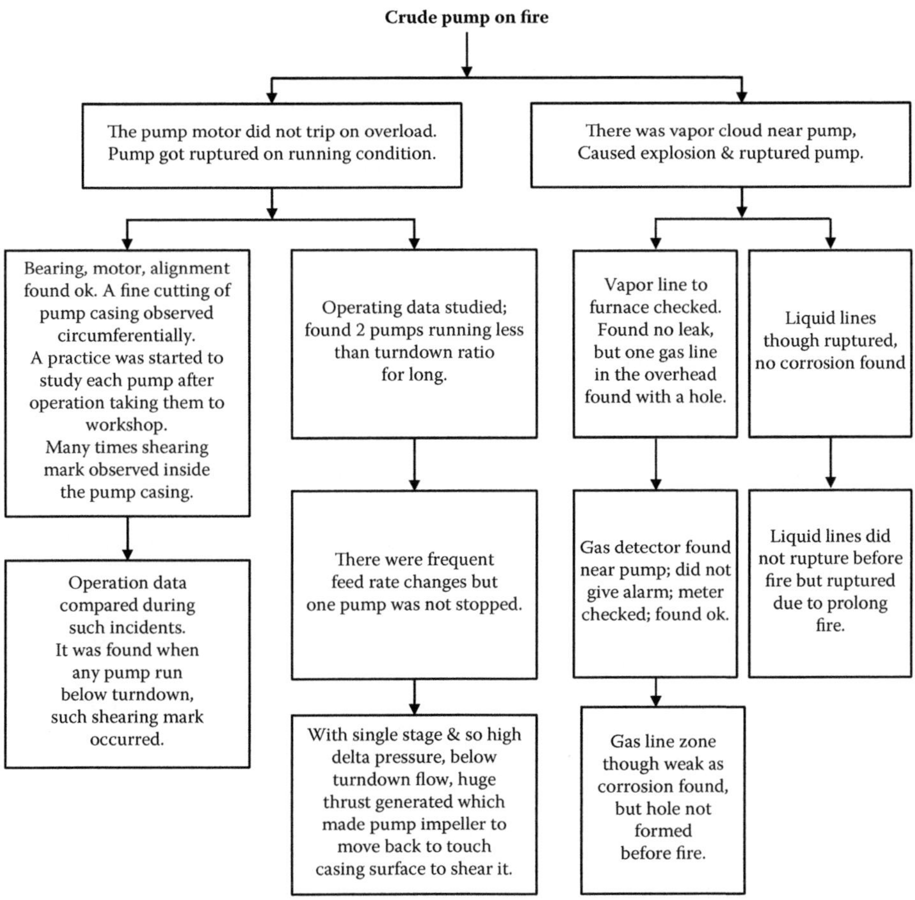

FLOW CHART 3.2

- Safety-hazardous and catastrophic events.
- Reliability-system unavailability.
- Performance-unintended functions.

b. **Disadvantages:**

- It is not a good tool for probability studies. RCA varieties are very diverse in nature and number and thus enough probability data may not be available.
- It is reverse to consequence analysis, and generic models are not applicable to predict the root cause of a possible anticipated failure.

3.2.2.2 Event Tree Analysis (ETA)

Event tree analysis (ETA) helps to identify the possible consequences/ events out of a basis event or number of events; i.e., it starts with basic event (s)/initial cause event(s). It determines the damage potential/impact of one or multiple causes of events. Here, the basic event or cause event is an accident event which is defined as a significant deviation from a normal condition that may lead to unwanted consequences like, gas leaks, explosions, fires, falls from a height and the like. The consequences may also depend on additional events and factors like:

- Whether the gas release is ignited or not.
- Whether people were present in the location or not.
- What the wind direction was during the event of gas leak and/or fire.
- Height of fall and whether it was freefall to land or equipment.

Many well-designed systems have one or multiple barriers which on implementation stop or reduce the consequences to a great extent. Hence, the probabilities of consequences will depend on whether the barriers are functioning or not. These barriers are called safety functions or protection layers, and may be technical or administrative ones.

ETA is an inductive procedure which shows all possible outcomes resulting out of an accidental initial event, taking into account whether safety barriers are functioning or not and also whether there are additional events and factors, such as those mentioned above, leading to the consequences.

By studying all relevant accident events using HAZID (hazard identification studies) and HAZOP (hazard and operability studies), and what if analysis ETA can provide to identify the potential accident scenario and sequences of consequences in a complex system. Through ETA, system weakness can be identified and probabilities of various outcomes can be determined as explained in Flow chart 3.3.

Similar exercises can be done to evaluate consequences probabilities on procedural lapses; i.e., on human error as a basic event.

Like the previous case of CDU crude pump failure shown in FTA earlier, here also, the same root cause can be considered as a basic event and then a logic tree may be constructed, but in this case downward. It can show the same various consequences as in the CDU case with various possibilities that can be categorized as less probable, probable and highly probable. Then, precautionary measures can be taken in the process of operation or design, as the case may be.

The steps to be followed in ETA are summarized below:

- Identify the initial/accident event that may result in an unwanted consequence.
- Identify the safety functions.

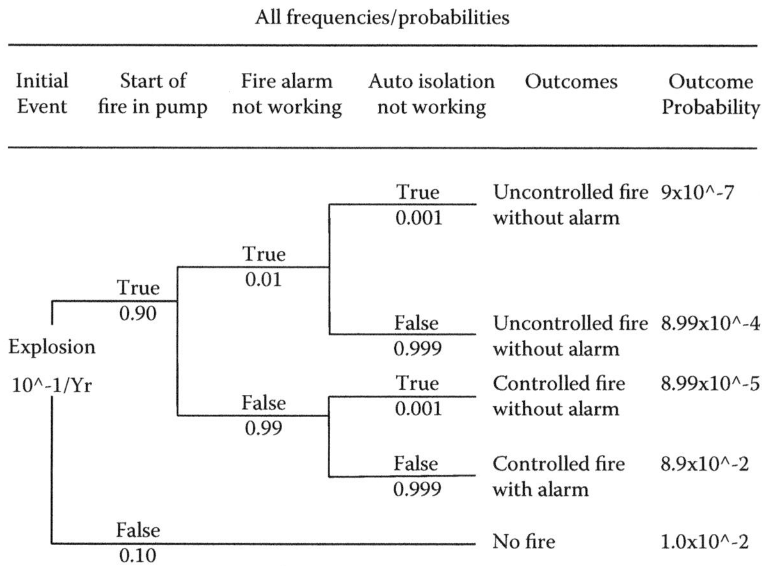

FLOW CHART 3.3

- Construct the event tree or logic diagram.
- Determine the potential consequences.
- Determine or use the database to know the frequency of basic accidental events
- Calculate the probabilities of the intermediate incident(s) and final consequences in the event tree.

The accidental event may be of three types:

- Equipment failure and/or process upsets
- Human error
- External reasons like an earthquake and/or cyclone.

Advantages and disadvantages of ETA:

a. **Advantages:**
    - It helps to visualize safety functions and sequence of activation of intermediate events.
    - It helps in making a good basis for incorporating safety features in the process by visualizing various possible safety functions.

b. **Disadvantages:**
    - It is not a tool to graphically represent the event tree.

- Only one initiating event can be studied in each analysis.
- There are easy to miss identifying inter-dependent subsystem(s).
- Analysis is not foolproof for common cause failures in quantitative analysis.

3.2.3 Consequence Modeling/QRA

Consequence (or impact) analysis is an important step in risk analysis. An accident begins with an accident, which usually results in a loss of containment of material.

Consequence modeling is a tool for calculation/estimation (numerical or graphical) of consequence/effect of a fire from inflammable, explosive and toxic material with respect to its impact on people, assets and/or other business sustainability parameters. It is one kind of risk assessment of a process plant and/or storage facilities handling hazardous liquids or chemicals which are extremely critical when the process is operated at high pressure and/or high temperature, moreover, when it is near highly populated areas. In consequence modeling, once the incident is defined, the source model is selected which provides a description of rate of discharge, total quantity discharged, duration of discharge and state of discharge. In consequence modeling, evaporation models are used to calculate the rate at which the liquid material becomes airborne which follows burning rate calculation from the pool and radiation heat flux from the flame to ambience. Next, a dispersion model is used to describe how the gas/vapor material is transported downward and dispersed in case of gas/vapor leakage from a source. For gas leakage of a jet, a jet fire model is used for fire hazard and loss estimation. There are explosion models also which are used for hazard and loss estimation from a concentrated gas mixture; i.e., a gas cloud in the atmosphere which on chance contact with an ignition source creates an explosion; i.e., a large fire with explosive sound. The models convert these incident-specific results into effects on people, structures/assets and the environment. However, environmental impacts are not covered in the modeling.

These models are used in quantitative risk analysis (QRA), a very common and widely used risk assessment methodology in a large or small project (grass root, revamp, and modification), which involves critical processes handling inflammable liquids or gases.

In QRA the impact of the scenarios is studied using available models like modeling of pool fire, modeling of VCE (vapor cloud explosion), BLEVE (boiling liquid expanding vapor explosion) and dispersion modeling.

Subsequently, risk estimation is done based on damage occurrence probability, as discussed earlier and impact assessment, as discussed in the present section.

QRA is most effective way to review and revise the current safety practices to make the process technologies more intrinsically and extrinsically safe.

Further follow-up exercises in QRA should also be carried out as follows:

- Uncertainty criteria study

- Risk control measures
- Risk acceptance

QRA can also be described in the form of Flow chart 3.4.

There are various mathematical models in use developed by various authors/institutions but for an organization it is simpler and better to use software models developed by different organizations which are popular and available in the market; some of these models are 'Aloha', 'Phast', and 'FRED'. Also, there are authorised agencies in each country that carry out these studies whenever called for; it is better to engage them for a project of a hazardous nature, but checking from time to time with minor changes in facilities or practice, the operating engineers can use these models to check the acceptability of the changes. Hence, to provide knowledge and understanding of the fundamentals, an attempt has been made to describe various consequences modeling in quantitative risk assessment (QRA) as follows.

3.2.3.1 Modeling of Pool Fires

Pool fire is a common type of fire, which can occur from an open roof tank or on tank roof rupture by corrosion/explosion or from a pool of open storage of inflammable liquid, say, open trench. It has been observed that characteristics of a pool fire depend on the pool diameter. There are various empirical models proposed by different authors, but The Netherland Organization for applied scientific research (TNO) introduced a model for estimation of radiating flux from a pool fire which seems to be more realistic for application in petroleum refineries; the model is commonly called as TNO

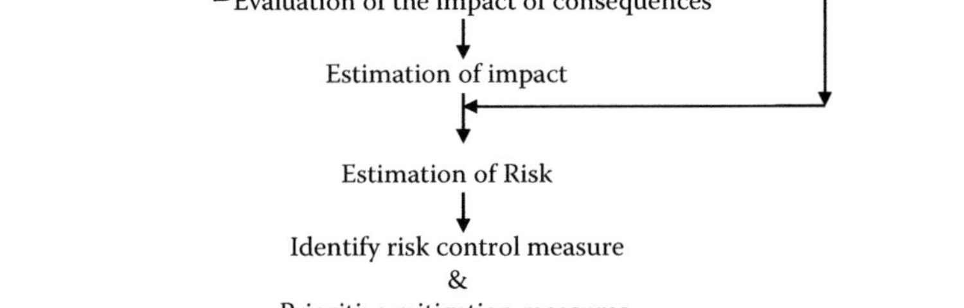

FLOW CHART 3.4

model (1 & 2). Basically, the heat flux model they have proposed is a derivative of Stiffen Boltzmann equation as given by equation 4 as follows:

$$H = \varsigma \ e_f (T^4 - t^4) \tag{4}$$

where H is the heat flux of radiation from flame, ς is emissivity of gray body radiation, e_f is Stiffen-Boltzmann constant, T is the temperature of flame and t is temperature of receiving surface, say, ambience.

For combustion process in fire, the classical model uses empirical equation to determine the burning rate, heat radiation and incident heat. For liquid having boiling point above ambient temperature, the rate of burning of liquid surface per unit area is given by equation 5 as follows:

$$\mathrm{dm/dt} = 0.001H_c/\left(C_p(T_b - T_a) + H_v\right) \tag{5}$$

where H_c = Heat of combustion (J/Kg), C_p = Specific heat at constant pressure (J/KgK), T_b = Boiling point, ^{0}K, T_a = Ambient temperature, ^{0}K, H_v = Heat of vaporization (J/Kg).

The above equation can be applied to evaluate the consequences of firing from a storage tank through lightning or other sources of ignition. Let us calculate the burning rate in gasoline tank firing as follows:

Ambient temperature, T_a	30° C, i.e., (30+273)= 303° K
Mid boiling point of Gasoline, T_b	115° C, i.e., (115+273)= 397° K
Specific Heat of Gasoline, C_p	2220 J/kg.K
Heat of vaporization of Gasoline, H_v	375000 J/kg
Heat of combustion of Gasoline, H_c	47000000 J/kg

So, from equation 5 above, the rate of burning per unit surface area from liquid surface under fire works out to:

$$\mathrm{dm/dt} = 0.001 \times 47000000/(2220(397 - 303) + 375000)$$
$$= 0.0805\mathrm{kg/sec.}$$

For liquid with a boiling point below ambient temperature, the expression is given by equation 6 as follows:

$$\mathrm{dm/dt} = 0.001H_c/H_v \tag{6}$$

In the petroleum refinery, the above equation applies to LPG or any other compressed liquid fire from a storage vessel or from a pipeline leak/rupture/release from a safety valve or drain valve because though the liquid is stored under ambient temperature, on release to atmosphere its temperature would go down below ambient temperature. Also, if the liquid instead is stored in cryogenic temperature to keep storage pressure atmospheric or near to it as presently is being followed in some industries, the same equation applies with following data.

- H_c for LPG = 460x105 Joules/kg
- H_v for LPG = 4.262x105 Joules/kg.

So, from equation 6, the rate of burning per unit surface area works out to:

$$dm/dt = 0.001 \times 460 \times 105/(4.262 \times 105)$$
$$= 0.108 kg/sec.$$

As a derivative of Stiffen=Boltzmann equation 4 above, There have been different correlations developed by various researchers for application area as discussed above and out of these, TNO model as discussed above is a good representation of heat flux from a petroleum storage tank/vessel fire and described as follows.

The total heat flux from a pool of radius, r (meter) is given by equation 7 as follows:

$$Q = ((\pi r^2 + 2\pi r H)(dm/dt)e_f H_c)/(72(dm/dt)^{0.61} + 1) \quad (7)$$

where Q = Total heat flux (W/m^2), H = Flame height (m), e_f = Efficiency factor. The efficiency factor of total combustion power is often quoted in the range of 0.15 to 0.35.

There are various empirical models to determine flame height, but the equation by G. Heskestad (3) appears to be simpler for application in real scenarios in the refineries. The properties of fire plume like flame height, temperature, velocity, concentration of combustions products and entrainment rate of air from surroundings have been discussed by G. Heskestad. He provided the correlation to find the flame height as given by equation 8 as follows:

$$H = 0.235Q^{2/5} - 1.02D \quad (8)$$

where D = Diameter of storage tank and Q = Total heat release by fire (kW/m^2).

From equation 7 & 8, it is evident that plume height, H is to be evaluated by iteration method.

The intensity of heat radiation at a distance r from the pool centre is given by equation 9 as follows:

$$Q_i = \tau Q/4\pi r^2 \quad (9)$$

TABLE 3.7

Effect of heat radiation to human and other lives

Intensity of radiation (kW/m^2)	Effect (human and inanimate)
1.6	No discomfort for long exposure
2.2	Threshold pain and no blister
4.2	First degree burn
8.3	Second degree burn
10.8	Third degree burn
15	Piloted ignition of wood
25	Spontaneous ignition of wood
4.0	Glass cracks
12	Plastic melts
19	Cable insulation degrades
37.5	Damage to process equipment
100	Steel structure fails

(Reference: AIChE/CCPS-Guidelines for chemical process quantitative risk analysis.)

where τ = Transmissivity of air path & Q = Heat flux ($W/m^{2)}$).

The effects of heat intensity of radiation to human burn and on others are shown in Table 3.7.

3.2.3.2 Modeling of Explosion

There are several types of explosions; namely detonation, vapor cloud explosion (VCE), and boiling liquid expanding vapor explosion (BLEVE). VCE and BLEVE are very common hazards; many times, petroleum industries have witnessed and suffered heavy damage due to these accidental events. For example, leaks from a LPG storage vessel, pipeline and pump in an offsite area connecting to a spark source outside the plant boundary. In the process plant leaks from pumps, heat exchangers handling flammable vapor and connecting to a nearby furnace cause VCE. Regarding BLEVE, it happened when there was water ingress into a storage tank from roof corrosion or from process plant along with the hydrocarbon. The temperature of the storage tank for heavy residual product or intermediates generally remains high. In such cases, storage temperature should be kept below 90°C or above 110°C, as temperature at or near 100°C would cause boiling of the entrained water inside the tank; and as a result, explosion would occur out of the sudden release of pressure; this is called BLEVE. It can occur even in pure hydrocarbon (also in the absence of water, if the storage temperature of flammable liquid reaches its boiling point). It can occur in various other scenarios, as discussed separately below.

TABLE 3.8

Damage potential of a blast w.r.t. its release pressure

Pressure (kPa)	Damage
0.14	Annoying noise (137 dB)
0.28	Loud noises (143 dB)
0.69	Breakage of windows under strain
1.03	Typical pressure of glass breakage
3.4-6.9	Large and small windows usually get shattered
4.8	Minor damage to house structures
6.9	Partial demolition of house
9.0	Steel frame slightly distorted
13.8	Partial collapse of house walls and roofs
17.2	50% destruction of brick works of house
34.5	Damage to wooden poles
34.5-48.2	Complete destruction of house
48.2	Loaded train wagons overturn
62.0	Loaded train boxes completely demolished
68.7	Total destruction of buildings, heavy machines

(Reference: AIChE/CCPS-Guidelines for chemical process quantitative risk analysis.)

In VCE or BLEVE, there are various consequences/effects of pressure release, as shown in Table 3.8.

A. **Modeling of vapor cloud explosion (VCE)**

When a large amount of flammable liquid or gas is rapidly released or slowly released for a longer duration unnoticed at less wind flow which helps in the accumulation of vapor instead of diluting it, a cloud forms which will mix with ambient air and very often form an explosive mixture. When the concentration of gas/vapor falls within a higher flammability/explosive limit (HEL) and lower limit (LEL) and the mixture comes once in contact with an igniting source, there will be huge explosion with fire. Similarly, the mixture on sudden release may generate heat by reverse Joule Thompson effect in case of hydrogen gas leak, which in turn causes a fire with an explosion, even without contact with any external igniting source. For hydrocarbon, the release can occur from a storage tank, Horton sphere of lighter hydrocarbon like LPG, propane, etc., process plant, transport vessel or pipeline. For hydrogen, release can occur from a storage vessel, pipeline or any equipment like heat exchanger or piping in process plant.

Following models are used for VCE modeling in the refineries:

- TNT equivalent model
- TNO multi energy model

These models provide empirical correlations based on limited field data and accident investigation studies. Out of these two models, TNO multi energy model is more effective in application in QRA as the method requires two parameters describing an explosion; namely a charge size (size of source of explosion) and charge strength (source overpressure). Also, computational fluid dynamics (CFD) can be used, which gives a framework to apply to the model.

TNT model, when applied to flammable vapor clouds, requires the explosion impact determined from past incidents.

The model developed by Baker, Strehlow and Tang is also used for flame speed evaluation from VCE.

There are various authors and organizations who consider different percentage of total fuel spill to combustion to consider in the model.

A1. *TNT equivalent model of VCE:*

The TNT equivalent of a flammable gas is the amount of the explosive TNT which causes the same impact of explosion as that of the subject flammable gas. Here, the model uses the simple heat energy equation considering an empirical explosion efficiency term. The TNT equivalent, E_m is given by equation 10 as follows:

$$E_m = e_f m H_c / H_{TNT} \tag{10}$$

where E_m = Equivalent mass of TNT (kg), e_f = Empirical explosion efficiency, m = Mass of flammable gas (kg), H_c = Heat of combustion of flammable gas (J/kg), H_{TNT} = Heat of combustion of TNT (J/kg).

A1.1 *Pressure of blast wave:*

The pressure wave effects are co-related as a function of scaled range. In VCE, generally a fraction of flammable material is considered to explode. This is generally in the range of 1–10%, which is called explosion efficiency. Hence, a scaled range is applicable here and is defined as the distance R by the cube root of TNT mass as shown by equation 11 as follows:

$$Z = R/W^{1/3} \tag{11}$$

where Z = Scaled distance in x-y graph with Z in x-axis and Scaled overpressure or impulse or arrival time or duration in y-axis but its dimension is $(m/Kg^{1/}3)$ where m is a length in meter and Kg is mass in Kg; R = Radial distance in meter from the surface of the fire ball, W = TNT equivalent (kg).

Using R and W, we can find out Z. From the graph, we can find out overpressure corresponding to Z.

A2. *Modeling of Boiling Liquid Expanding Vapor Explosion (BLEVE):*

BLEVE has been explained earlier. In addition to the area mentioned earlier, BLEVE can occur in the transportation of hazardous material as well. BLEVEs are important, especially due to their severity and the fact they involve simultaneously diverse effects which can cover large areas, overpressure, thermal radiation and projectile effect of the fragments generated in the blast. To explain the severity of explosion with respect to BLEVE, it can be mentioned that LPG liquid expands about 270 times and in the case of propane it is even higher when it boils to the vapor stage from the liquid stage; thus creating a huge expansion to cause the explosion.

In some BLEVE cases caused by exposure to fire, the flammable liquid can go to BLEVE stage either by the heating of its container in fire or if the container in fire ruptures, causing pressure reduction of the flammable liquid inside, which in turn leads to its boiling point to come down to reach the BLEVE stage. Hence, the major hazards of BLEVE are by water contamination, thermal radiation, and overpressure or sometimes by the velocity of any fragment that ruptures the storage vessel on explosion to the distance of hitting any object or human being. Following are some models appearing to be realistic for application in petroleum refineries.

A2.1 *Pressure of blast wave due to BLEVE:*

Pressure effect from the blast from a distance of a storage vessel is given by Baker et el (4) and Prugh (5) as equation 12 & 13 as follows:

$$W = 3.662 \times 10^{-6} V (P_1/P_0) R_g T_0 \ln(P_1/P_2) \qquad (12)$$

where W = Equivalent energy (kg TNT), V = volume of compressed gas (m^3), P_1 = pressure of the compressed gas (N/m^2), P_2 = final pressure of the expanded gas (N/m^2), P_0 = standard pressure (N/m^2), R_g = gas constant (J/Kg.mole.^{0}K), T_0 = standard temperature. For storage of any hydrocarbon, all the above operating conditions like V, P_1, P_0, and T_0 are known; by collecting TNT equivalent (W) and R_0 from the property tables of the hydrocarbon, we can calculate the value of P_2, i.e., final pressure of expanded gas, i.e., blast effect to know the impact of explosion. Similarly from the following equation 13 also, we can calculate burst pressure as follows:

$$P_b = P_s(1 - 3.5(\gamma - 1)(P_s - 1)/\sqrt{(\gamma T/M(1 + 5.9P_s)))}^{-2\gamma/(\gamma-1)} \qquad (13)$$

where P_s = pressure at the surface of the vessel (bar.abs), P_b = burst pressure of the vessel (bar.abs), γ = heat capacity of expanding gas, M = molecular weight of expanding gas (gm.mole), T = absolute temperature of the expanding gas (^{0}K).

Here, value for the distance, R from the explosion centre is obtained from equation 11, and equivalent energy of TNT, W is calculated from equation 12.

Distance from centre of the pressurized container to its surface is to be subtracted from the distance R to produce the virtual distance to be added to distance for shock wave calculation.

The overpressure at any distance is determined by adding virtual distance to actual distance, and then using the distance to determine the scaled distance, Z. Then monograph developed by AIChE/CCPS is to be used to determine the overpressure.

A2.2 *Model on radiation received by target:*

The radiation received by receptor (for the duration of BLEVE incident) is given by CCPS as equation 14 as follows:

$$E_r = \tau_a E F_{21} \tag{14}$$

where E_r = Emissive radiation flux received by a receptor (W/m^2), τ_a = Transmissivity (dimensionless), E = Surface emitted radiation flux (W/m^2), F_{21} = View factor (dimensionless).

Different researchers also developed correlations to estimate the surface heat flux from the radiation of the flame generated by heat of combustion on BLEVE fire as named below:pt

- Roberts (6), Hymes (7) and CCPS (8)
- Pieterson and Huerta (9) and TNO (10)

Empirical co-relations also developed by CCPS for determining the diameter of a fire ball, distance between fire ball and the object in the ground and duration of the BLEVE are shown in equation 15, 16 & 17 as follows:

$$D = 5.8m^{1/3} \tag{15}$$

$$L = 0.75D \tag{16}$$

$$T = 2.6m^{1/6} \tag{17}$$

where m= mass of the flammable material, D= diameter of the fire ball, L= distance of the fire ball up to the ground, and T= duration of BLEVE.

There are also correlations on velocity of fragments generated in the BLEVE.

Baker et el (4) and Brown (11) developed correlation for velocity of fragments from blast of cylindrical-spherical vessel in the BLEVE.

LPG bottling plants are common not only in petroleum refineries but there are also many standalone LPG bottling plants in various places across all countries. Hence, utmost care should be taken with respect to adequate safety measures to avoid the terrible consequence of VCE and BLEVE, as discussed above. LPG cylinders can produce about 2 to 4 fragments to generate the projectile to hit any object.

In LPG storage bullets or spheres, working pressure remains at about 60 psig and the storage vessels are generally designed at a pressure of 5 to 6 times of working pressure; even at that design pressure, vessels get ruptured during BLEVE.

3.2.3.3 Dispersion Modeling

Dispersion means moving and spreading through diffusion in the medium of air or other areas as the case may be, but when discussed with respect to a hazardous incident, the medium is referred to as ambient air. If the dispersing vapor is heavier than air, then the vapor cloud moves in a downwind direction and spreads or disperses through a diffusion process in a crosswind. But if the gas is lighter than air, then it moves upward with some intermix in downward direction. The second one is less risky; i.e., if the gas is lighter than air, but the first one is very hazardous and often happens in petroleum industries. For example, many times installation flares have been found extinguished. In such a situation, gas cloud accumulates in the ground level and it has been reported from a nearby village school where students reported an odd smell in the air; actually the problem occurred in a refinery sour gas incinerator flare which got extinguished; as there is H_2S gas in the sour gas, a slight escape of H_2S would cause a pungent smell like that from a domestic LPG cylinder in a household kitchen. In such a situation, people detect the issue at an early stage by smelling it before the gas reaches a concentration level for explosion, as in the case of the school students mentioned above. But in a refinery main flare, the smell may not be there as is true of pilot gas, used for keeping flare on; and main flare gas may be sweet; i.e., without H_2S to avoid piping corrosion. Hence, people have to be watchful on keeping a pilot flare on; also, in case of extinguishing, if any equipment safety valves pop, there may be huge gas dispersion to the atmosphere and accumulation in the ground, allowing no time to take remedial measures.

Dispersion modeling is a mathematical simulation governing the transport, dispersion and transformation to form a vapor cloud at some location at the ground level. The model helps to estimate the area affected and average vapor concentration in the cloud. The inputs in the calculation are: rate of the gas released, release source diameter, wind speed, ambient temperature, humidity, and type of the location in the ground, i.e., whether there is less restriction air flow to help dilution of the vapor release below a hazardous concentration level. The model is also used to estimate air pollutant concentration in the localities.

Fundamentally, the well known Gaussian dispersion equation describes the behavior of neutrally buoyant gas released in the wind direction. It is a time average model and thus gives more a practical estimate. Instead of detailing the equation, which is available to the readers even through the web version of the literature, it is worth describing the equation through pictorial presentation as shown in Figure 3.2.

It is recommended to use a software model on dispersion for industrial/commercial purposes. The software, CHARM is widely used for this purpose. The software not only provides the estimate of dispersion but also facilitates contingency planning in such an emergency, and thus helps to meet the organization's environment management program. There are two versions of CHARM; the first version assumes a single source in flat terrain but the second version covers multiple sources that are also in complex terrain but require more

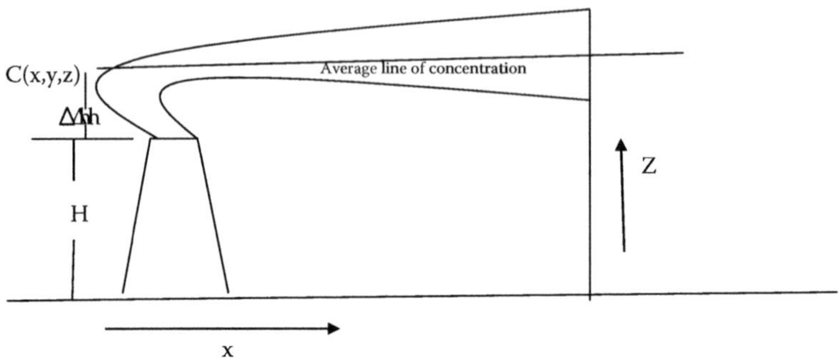

FIGURE 3.2
Vapour cloud dispersion diagram on toxic gas release from a height. C(x,y,z) is the concentration of stack gas plume combined in all three directions, x, y and z; x and y are in horizontal plane and z is in vertical direction as shown; H is the height of the stack and Delta h is the height of the average central line of plume concentration from stack top level, Q is the quantity of source stack gas emitted from stack top which is not shown above but used in the dispersion equation.

input data. A Web version is available for the first version and is suitable for beginners to try. For commercial use, it is recommended to line up an authorized agency to do the exercise, as there are more models available in the market, say, 'Aloha,' and commercial agencies use the right model for their client(s).

3.2.4 Risk and Its Categorizations

Risk is a product of frequency and consequence. Different levels of Risk defined as per IEC standards are shown in Table 3.9 and Table 3.10 as follows.

3.2.4.1 Risk Calculation

Risk is expressed as the product of frequency of an event and the magnitude of consequences. In this category, only the property and human resource

TABLE 3.9

Following categorizations w.r.t. risk frequencies and risks severity are being followed internationally

Risk levels based on frequency		
Risk level	**Descriptor**	**Frequency of occurrence**
5	Frequent	One per year
4	Probable	One per 10 years
3	Occasional	One per 100 years
2	Remote/Rare	One per 1000 years
1	Improbable	One per 10,000 years

TABLE 3.10

Risk levels based on severity

Risk level	Descriptor	Potential consequences
5	Catastrophic	Multiple deaths
4	Severe	Death
3	Serious	Lost time accident
2	Minor	Medical treatment
1	Negligible	No injury

Note: In the industries, generally, if an employee/work force report after medical treatment within 72 hours of injury, then lost time accident is considered zero and if such an accident is reported after that period, it will be treated as injury. Such risk level belongs to 'Minor injury' category, i.e., risk level 2.

damage, i.e., human loss and economic loss are considered. Risk criteria, which are the acceptable levels of risk that can be tolerated under a particular situation for industrial installations by a few countries, are shown as an annexure in the respective code.

The above risk assessment study should be carried out by an authorized agency.

While studying the process to be used for operation, it is a general practice to subdivide the whole process into a number of subprocesses, and risk analysis should be carried out for each subprocess, as explained above. In each subprocess, equipment-wise risk analysis also should be done.

3.3 HAZOP Study

HAZOP studies stands for hazard and operability studies of any project, production process of the industries. HAZOP studies cover hazard identification only, i.e., it is qualitative in nature but when added with risk mitigation measures it comes under quantitative risk analysis (QRA).

Though qualitative in nature, the techniques followed in HAZOP studies help in identifying all minor and major hazards of a very complex system comprising many processes like distillation, separation, reaction, high pressure, high temperature scenario, variation of pressure zones, etc., unlike conventional hazard identification techniques. Hence, HAZOP studies are very useful and followed in petroleum and other complex process industries. The techniques followed in HAZOP studies are somewhat different than analyses that are followed in processes but restricted to simple processes. Here, the techniques use many keywords like More, Less, Reverse, No in "flow" scenario; More, Less in "pressure" scenario; More, Less in "temperature" scenario; High, Low, No in "level" scenario; High, low in "product qualities" scenario; Part of in "composition in a system" which is "missing or more/less"; Other than in "unwanted material, explosive pressure, unwanted reaction(s)", and so on. These guidewords are used to understand the deviation of the process conditions from normal levels.

HAZOP studies can be carried out at all levels of activities in the industries as follows:

- Process development
- Defining project standard
- Process design
- Engineering and procurement
- Project execution
- Commissioning
- Normal operation
- Normal startup, shutdown
- Emergency shutdowns

As any project has many levels like mentioned above, degrees of HAZOP studies also would be different at various levels.

In the initial levels like process development and defining project standards, the following hazard studies are carried out in the form of reasoning and what if analysis:

- Broad identification of the hazards associated with the process(s).
- Identification of environmental issues associated with the site, population and demography.
- Collecting a history of previous hazards associated with the processes being contemplated and on the site such as earthquakes, cyclones, water availability crises, threats on possible emergency responses, demographics, etc.

Note: Due to above, after finalization of the production process configuration and site, an environment impact assessment (EIA) study is undertaken by the industries. Industries, generally line up an authorized agency to carry out the EIA study.

For process design, engineering and procurement, the second stage of hazard studies are carried out as follows. Again, they are based on historical data and information, along with government regulations:

- Examine the plant items critically with respect to associated design hazard, logistic hazard and installation hazard in the selected site.
- Identify the alternatives wherever applicable; include and review its impact on the EIA already carried out.

For process design like finalization of process flow diagrams (PFDs), piping & instrumentation diagrams (P&IDs), unit plot plan, startup/shutdown procedures, and also for incorporating any change in process scheme, operation philosophies and startup (including commissioning)/shutdown procedures, HAZOP studies are carried out as the third stage of hazard studies with the following objectives:

- Identifying all possible hazards in accepting the PFDs, P&IDs, startup (including commissioning)/shutdown procedures, unit plot plan in the design as prepared and also in incorporating intended modification/changes during operation.
- Identifying/evaluation of possible mitigation measures.

Note: when risk evaluation is done on consequences/risk quantum with and without executing mitigation measures, then the above HAZOP studies become a part of QRA studies.

Even without QRA studies, industries execute the identified mitigation measures or reject some of the recommended mitigation measures when the hazards are clearly understood at the organization level. This is called the fourth stage of the project.

When mitigation measures are taken, that means again there has been some change incorporated into the overall project. Hence, the consequences of these changes are to be reviewed by reworking the original QRA if not done while implementing the mitigation measures. Also the effect on the original EIA should be checked when the project safety evaluation comes at the closure stage, which is the fifth and final stage of hazard analysis.

HAZOP studies can be explained as shown in Flow chart 3.5.

Various guide words in selecting deviation in HAZOP studies

Guide words (keywords) used in HAZOP studies are shown along with their interpretation in Table 3.11.

HAZOP studies worksheet preparation

Before conducting HAZOP studies, committees or team should be formed comprising employees experienced in the domain with responsibilities fixed with respect to completion of the studies and taking a lead role in executing or coordinating execution of the mitigation measures identified during HAZOP studies. Sometimes, for a big project, external agencies are lined up to carry out the HAZOP studies with fixed responsibilities from the owner's side to coordinate with the agency(s) and ensure approval of the study report by a competent authority both from the agency and from the side of the owner.

To avoid missing to check any part of the P&IDs and process, the checklist should be in hand for reference.

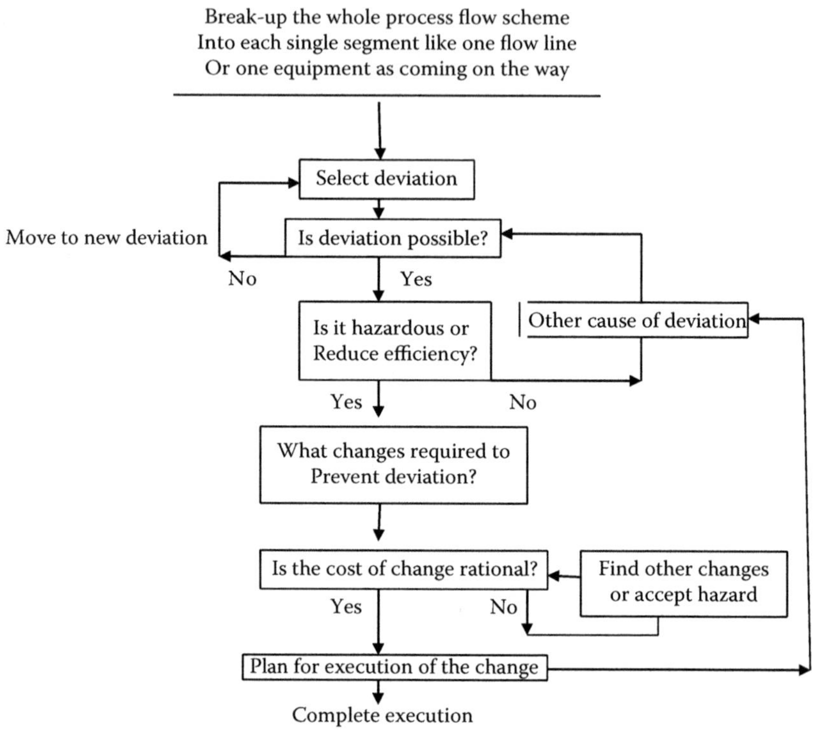

FLOW CHART 3.5

TABLE 3.11

Applicability areas of guide words

Guide word	Meaning / interpretation
No or Not	Flow, Level, Alarm, Indication, Trip, Reaction, Liquid 'X', 'Y' / Vapor 'X', 'Y'
More or High	Flow, Level, Volume, Pressure, Temperature, Exotherm / Endotherm
Less or Low	Flow, Level, Volume, Pressure, Temperature, Exotherm / Endotherm
As well as	Low flow as well as low temperature, No alarm as well as no indication, and so on from above
Part of	Flow, Reaction, Volume, Concentration
And	Liquid 'X' and Liquid 'Y', Liquid 'X' and Vapor "X', etc.
Or	Liquid 'X' or Liquid 'Y', Liquid 'X' or Vapor "X', etc.
Reverse	Flow, Pressure
Before/After	Feed flow cut off before fuel flow cut off or vice versa
Early/Late	Early control valve closing / opening w.r.t. flow, level, pressure, temperature or vice versa

TABLE 3.12

HAZOP study worksheet

Sl. No.	Guide word	Deviation	Cause(s)	Consequence	Mitigation measure(s) identified	Signed by
1	No	Flow	-Pump fail -upstream (u/s) / downstream (d/s) Control valve closed -u/s or d/s Line choked -u/s or d/s check valve in reverse direction	-Feed to d/s heater cut off & furnace o/l temp. rise -u/s vessel, if any, level will rise -production interrupted	-Provide fuel auto cut off to heater on feed failure -provide level control valve at inlet of u/s vessel -enhance line heat tracing if it is poor	

Note: The above example is given with only one guide word and that too with only one deviation. So, for the same guide word, 'No', there will be other deviations like level, alarm, etc., as mentioned earlier and the above exercise is to be carried out for all these deviations also. Similarly, there are other guide words as mentioned earlier and hence, the above exercise can also be carried for these as well; i.e., the number of rows in the above table would increase.

Worksheet shown in Table 3.12 should also contain the following:

- Project title:
- P&ID reference no:
- Revision No. with date:
- Study line no:
- System description:
- List of team members:
- Chaired by:

HAZOP case study

A desalted and preheated crude oil is subsequently routed to a fuel fired heater for further temperature raising necessary before sending to a crude distillation column in a refinery's crude distillation unit (CDU). Heater inlet temperature in CDU in the industry varies between 260° C to 280° C depending upon the heat exchange efficiency in the preheat train. The object of the fuel fired heater is to ensure fuel oil and/or fuel gas flow through respective burners and fire inside the heater box using atmospheric air heated in air pre-heater under balanced draft or using cold natural air under natural draft for combustion to raise the

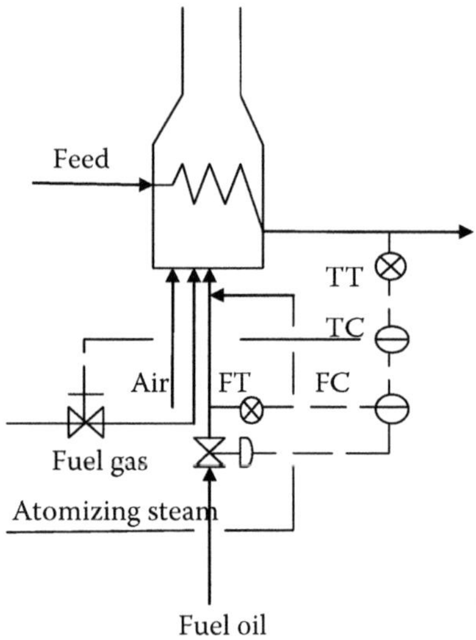

FIGURE 3.3
Legend: FT: Flow transmitter, FC: Flow controller, TC: Temperature controller, TT: Temperature transmitter. *Note:* In actual, there is also a pressure control valve in atomizing steam line (not shown) which would open as per a differential pressure controller (not shown) between atomizing steam and fuel oil line from the pressure impulse line and pressure transmitter (PT) at the d/s of the respective control valves (not shown) to maintain atomizing steam pressure higher by - 1.5 to 2 bar at burner - (based on burner design data).

heater outlet temperature of the crude oil up to about 370° C which then flows to the distillation column for flash vaporization of crude oil in order to get various distillate products/intermediates from the distillation column. To begin to understand the HAZOP case study, let us take a small section of heating the crude oil in the fired heater, as shown schematically in Figure 3.3. The HZAOP analysis on the above scheme is based on a section of PFDs or P&IDs with a bunch of lines and control elements and is shown in Table 3.13 using the guide words mentioned earlier.

If the HAZOP team requires deciding whether a part of mitigation measures can be avoided or not, then risk analysis should be done on the effect of not implementing any mitigation measure. This analysis outcome should be reproduced in the form of a risk matrix where risks are shown as very high, high, medium and low as shown in Table 3.14.

3.4 Adherence to Standards on Mechanical Engineering and Installation Criteria

All standards must be followed while using the design/engineering calculations as followed by technology houses and engineering houses. For mechanical designs, following are the standards to be followed:

ASTM (American Society for Testing and Material): the standards include test procedures for determining or verifying characteristics as chemical composition, measuring performance. The standards cover refined material as steel and basic products as machinery and fabricated equipment. ASTM standards are published in a set of 67 volumes in 16 sections.

TABLE 3.13

HAZOP case study worksheet

Guide word	Deviation	Cause	Consequences	Mitigation
No	Flow of feed	-Feed pump tripped.	-Heater outlet temp. Shoot up & heater tube may coke up.	-Auto cut off fuel to provide. -High temp. & low flow alarm to provide.
Low	Flow of feed	-Feed pump strainer choked.-Low level in u/s vessel.	-Heater outlet temp. shoot up & heater tube may coke up.	-Auto cut off fuel to provide. -High temp. & low flow alarm to provide.
High	Flow of feed	-Feed pump discharge or heater inlet control valve not working.-Feed pump u/s vessel level control not working.	-Feed pump may trip-Heater o/l temp. may shoot up-Feed pump u/s vessel level may become low so that pump lose suction & low feed flow before pump trip.	-Auto cut off fuel to provide. -High temp. & low flow alarm to provide. -Control valve to be rectified by isolating & bypassing it before trip using low flow/high temp. alarm.
Low	Temperature	-Feed flow high as flow control not working.-Fuel oil / fuel gas flow, control not working.-Fuel oil / gas burners malfunctioning.	-Inadequate distillate recovery from the distillation column.-Quality of products going off.	-Low temp., high feed flow alarm to provide so that timely actions can be taken.-Burners should be cleaned and fuel flow control to be checked & rectified.
High	Temperature	-Low feed flow.-High fuel flow.	-Heater outlet temp. shoot up & heater tube may coke up.	-Auto cut off fuel to provide.-High temp. & low flow alarm to provide.
Low	Fuel oil flow	-Fuel oil pump tripped or fuel oil vs. atomizing steam diff. control valve not working.-Fuel oil burners choked.	-Heater outlet temp. to reduce & affect distillate recovery & quality.	-Feed should reduce on auto temp. control or gas flow to increase to hold temp.-Fuel oil

(Continued)

TABLE 3.13 (Continued)

Guide word	Deviation	Cause	Consequences	Mitigation
				low pressure alarm to provide to manage with gas burners till that time.
No/ Low	Atomizing steam pressure	-Fuel oil vs. atomizing steam diff pr. Control valve not working.--Steam header pr. Come down	-Heater outlet temperature comes down resulting in low distillate recovery & quality.	-Alarm for both low diff. press. & steam header press. Reqd. -Feed flow to reduce on auto temp. control. -Fuel gas flow to increase on auto temp. control.
As well as	Both fuel oil & fuel gas flow low	-Fuel oil pump trip as well as fuel gas press low.	-Heater outlet temperature comes down resulting in low distillate recovery & quality.	-Auto temp control to reduce the feed flow.-Low fuel oil & fuel gas flow alarm to provide.

Note: Similarly, cases with low heater box draft caused by stack damper not opening or due to less combustion air flow, cases with low oxygen in heater box and high heater tubes' skin temperature cases are to be worked out by expanding the above table.

TABLE 3.14

Hazard severity vs. likelihood

	Likelihood			
Severity	**Frequent**	**Probable**	**Occasional**	**Remote**
Major	Very high	Very high	Very high	High
Moderate	Very high	High	High	Medium
Small	High	Medium	Low	Low
Insignificant	Medium	Low	Low	Low

ASME (American Society for Mechanical Engineers): It is a leading organization in the world developing codes and standards. The work of the society is performed by its member-elected board of governors and through its five councils, 44 boards and hundreds of committees in 13 regions throughout the world. It is a 120,000 member professional organization focussed on technical, educational, and research issues of the engineering and technical community. ASME conducts one of the largest technical publishing operations and holds numerous technical conferences worldwide and offers hundreds of professional development courses each year. ASME sets internationally recognized industrial and manufacturing codes and standards that enhance public safety. The work of the society is performed by its member-elected board of governors and through its five councils, 44 boards and hundreds of committees in 13 regions throughout the world. Advancing the science and practice of mechanical engineering is the responsibility of the society's 37 technical divisions and subdivisions, which span a vast variety of disciplines, technologies and industries.

ANSI (American National standard Institute): It is a forum for development of American national standards. It is a private, non-profit organization that administers and co-ordinates the US voluntary standardization and conformity assessment system. Safety is the basic object of the standards developed by ANSI. The ANSI standards include prohibition for practices considered unsafe. ANSI provides a forum for development of American national standards from organizations such as ASME, NFPA (National Fire Protection Association) and more, and services as co-ordination point for national distribution of international standards issued from organizations such as ISO (International Organization for Standardarization), DIN (Deusches Institut fur Normung eV), IEC (International Electro Technical Commission) and others.

API (American Petroleum Institute): It represents more than 400 members involved in the oil and natural gas industries. Over 900 API standards serve as the basis for API quality programs covering production material and lubricants, and certification programs for storage tanks, pressure vessels, and piping inspectors. API publish recommended practices, research reports, and specifications on pipelines, valves, off-shore structures, oil spill response procedures, environmental protection, exploration, facility management and much more. API membership work is organized in a) The API upstream segment, b) the API marine segment, c) the API downstream segment, and d) the API pipeline segment.

NACE (National Association of Corrosion Engineers): it was established in 1943 by eleven corrosion engineers in the pipeline industry. NACE standard practices are methods of selection, design, installation, or operation of a material or system when corrosion is a factor.

ASME standards for mechanical systems

- ASME SEC I: Rules for construction of power boilers.
- ASME SEC II Material part A: Ferrous material specifications.
- ASME SEC II Material part B: Non ferrous material specifications.
- ASME SEC II part C: Specifications for welding rods, electrodes and filler metals.
- ASME SEC II part D: Properties.

- ASME SEC III: Rules for construction of nuclear plant components.
- ASME SEC III Subsection NCA: General requirements for division 1 and 2.
- ASME SEC III Division 1 subsection NB for Class 1 components, NC for class 2 components, ND for class 3 components, NE for class MC components, NF for supports, NG for core support structures, NH for components in elevated temperature service.
- ASME SEC III Division 2: Code for concrete reactor vessels and containments.
- ASME SEC IV: Heating boilers.
- ASME SEC V: Non-destructive examination.
- ASME SEC VI: Recommended rules for the care and operation of heating boilers.
- ASME SEC VII: Recommended guidelines for the care of power boilers.
- ASME SEC VIII: Rules for construction of pressure vessels in Division 1 and alternate rules in Division 2.
- ASME SEC IX: Welding and brazing qualifications.
- ASME SEC X: Fiber-reinforced plastic pressure vessels.
- ASME SEC XI: Rules for in service inspection of nuclear power plant components.
- ASME B31: Code for pressure piping, developed by ASME, covers power piping, fuel gas piping, process piping, pipeline transportation systems for liquid hydrocarbons and other liquids, refrigeration piping and heat transfer components, building services piping and slurry transportation piping systems. ASME B31 was earlier known as ANSI B31. ASME B36.10M (table 2 and 4 of ASME tables) and, ASME B36.19M (table 1 of ASME tables) cover piping dimension standards for CS and SS.
- API RP 520: sizing selection and installation of pressure relieving devices in refineries.
- API 610: Centrifugal pumps for petroleum, heavy duty chemical and gas industry services.
- ANSI/NEMA SM23: Steam turbines for mechanical drive services.
- API 617: Centrifugal compressors for petroleum, chemical and gas industry services.
- API 682: Shaft sealing system of centrifugal and rotary pumps.
- EJMA codes: Codes developed by Expansion Joints Manufacturer's Association.
- NACE MR 0175: Sulfide stress cracking resistant metallic materials for oilfield equipment.

- NACE TM 0177: Laboratory testing of metals for resistance to sulphide stress cracking in H2S environment.
- IS 1239: Mild steel ERW tubes for ordinary use in water, gas, steam and air lines.

OISD standards

Many countries have their own regulations and guidelines based on the above codes as well as their own respective operation and maintenance experiences and research. Hence, they have implemented their own standards, which are to be followed over and above the international standards mentioned above. Different OISD standards in India for oil and gas installations are given as follows with respect to mechanical engineering, along with general features required for the installation:

- OISD 118: Layout for oil and gas installations specifically valid for India.
- OISD STDs 106, 109, 111, 118, 163, 178: Process & Installation.
- OISD STD 166, 196, 197, 200: Environment protection.
- OISD STD 105: Work permit system.
- OISD STD 119: Inspection of pumps.
- OISD STD 120: Inspection of compressors.
- OISD STD 121: Inspection of turbines and diesel engines.
- OISD STD 122: Inspection of fans, blowers, gear boxes and agitators.
- OISD STD 123: Inspection of rotating equipment components.
- OISD RP 124: Predictive maintenance practices.
- OISD STD 125: Inspection and maintenance of mechanical seals.
- OISD RP 126: Specific practices for installation and maintenance rotating equipments.
- OISD STD 128: Inspection of unfired pressure vessels.
- OISD STD 129: Inspection of storage tanks.
- OISD STD 130: Inspection of pipes, valves and fittings.
- OISD STD 131: Inspection of boilers.
- OISD STD 132: Inspection of pressure relieving systems.
- OISD STD 133: Inspection of fired boilers.
- OISD STD 134: Inspection heat exchangers.
- OISD STD 135: Inspection of loading and unloading hoses for petroleum products.
- OISD STD 171: Preservation of idle static and rotary mechanical equipments.

- OISD STD 195: Safety in design, operation, Inspection and maintenance of hydrocarbon gas compression stations and terminals.
- OISD STD 203: Operation, maintenance and inspection of hoisting equipments.
- OISD RP 205: Crane operation, maintenance and testing for upstream industries.

Note: While the above OISD guidelines are directive principles, the same have become mandatory to be followed to get statutory approval from PESO (Petroleum and Explosive Safety Organization), an organization in India set up long before OISD, to regulate safety in petroleum and chemical installations. In addition, for operation in the upstream, DGMS, India guidelines developed for mines operation are to be followed. For products, BIS specifications and for ATF (Aviation Turbine Fuel), additionally DGCA approval is needed in India.

Act for steam generators/boilers (IBR)

For steam boilers, Indian industries have additionally followed the Indian Boiler Regulation (IBR) which was created on 15th September 1950 in exercise of the powers conferred by section 28 & 29 of Indian Boilers Act which was formed 23rd February 1923 to consolidate and amend the law relating to steam boilers. Act contains 741 total regulations. Steam boilers under IBR means any closed vessel exceeding 22.5 litres in capacity and which is used expressively for generating steam under pressure and includes any mounting or other fitting attached to such vessel which is wholly or partly under pressure when steam is shut off. Regarding steam pipes, IBR steam pipe means any pipe through which steam passes from a boiler to a prime mover or other user or both if pressure at which steam passes through such pipes exceeds 3.5 Kg/Cm2 above atmospheric pressure or such pipe exceeds 254 mm in internal diameter and includes in either case any connected fitting of a steam pipe.

3.5 Adherence to Standards with Advanced Instrumentation and Automation

Process should be integrated with a safety instrumented system (SIS). Process instrumentation has undergone rapid changes in the recent past and today the focus is on providing automation with the highest safety standards. These safety standards are derived from internationally accepted codes and practices. Regulatory bodies like Instrument Society of America (ISA), American National Standards Institute (INSI), Institute of Electrical & Electronics Engineers (IEEE), International Electro-technical Commission (IEC), European Committee for Electro-technical Standardization (CENELEC), Oil Industry Safety Directorate (OISD), Bureau of Indian Standards (BIS) and so on formulate these codes and standards to ensure safe operation.

3.5.1 Codes on Instrumentation

Some of the common instrumentation codes and standards used in the process industries as laid down by the above bodies are enumerated as follows:

- OISD 152 & 153: This standard on maintenance and inspection of safety instrumentation in the hydrocarbon industry provides the guidance for removing obsolescence, standardizing and upgrading existing practices and standards to ensure safe operation.
- IEC 61511: This technical standard sets out practices in the engineering system that ensures the safety of an industrial process through the use of instrumentation. Such systems are referred to as Safety Instrumented System (SIS). The title of the standard is Functional Safety-Safety Instrumented System for the process industry sector. The standard also recommends the practice for evaluating Safety Integrity Level (SIL).
- IEC 60079: This standard which is also referred to as EN 60079 (EN stands for European Standards Compliance) or ISA-12.04.01-2004 for the guidance for selection of enclosures for housing electrical and electronic apparatus for use in the explosive gas atmosphere. The standard also provides the guidelines for use of devices in intrinsically safe circuits.
- ANSI/ISA-91.00.01-2001: This standard guides identification of Emergency Shutdown Systems and Controls that are critical for maintaining safety in process industries.
- ANSI/ISA-TR99.00.01.2007: This ISA Transaction Report provides the guidelines for security technologies for industrial automations and control systems.
- ANSI/ISA-18.2.2007: This provides guidelines on the management of alarm systems in the process industries.
- ANSI/ISA-61010-1 (82.02.01)-2004: This provides direction on safety requirements for electrical equipment, measurement, control and laboratories.
- ISA-TR84.00.07-2010: This provides guidance for evaluation of fire and gas detection system effectiveness.
- ISA-TR98.00.02.2007: This provides skill standards for instrumentation technicians.

3.5.2 Safety Instrumented System (SIS)

Age-old instrumentation systems are no longer safe for operation. New international standards for safety, like IEC 61508 and IEC 61511, are prompting a reexamination of safety practices. Planning is required to meet increased regulatory requirements across the globe. To develop this plan, one needs to effectively perform hazard identification, hazard analysis, and risk assessment, as mentioned earlier.

IEC 61508 is used by the suppliers of safety-related equipment; it also defines a set of standards for functional safety of electrical/electronic/programmable electronic safety related systems. So, a supplier should be IEC 61508 certified; also it needs to manufacture the equipment in accordance with IEC 61511. In IEC 61511, formally collected best practices in safety applications have been defined.

To have the whole process in keeping with an advanced safety system—i.e., to be a smart process—the basic process control system (BPCS) should be clubbed with SIS. A SIS contains three parts; namely sensors, logic solvers, and final control elements (valves) to increase safety. Regarding malfunction of the system, it is a statistically established fact that in 42% of cases sensor malfunction, 8% of cases logic solver malfunction and in 50% of cases valve malfunctions are responsible. Sensors for pressure, temperature, flow and level play an important role in risk reduction strategy. It is important to consider improvement in measurement technology as well as installation and maintenance practices. The next most important step in SIS is to equip the final elements with digital valve controllers that provide the diagnostics to extend the proof test interval, while delivering higher reliability and safety. Also, bulky logic solvers and multiplexers can be replaced with state-of-the-art logic solvers that support digital communications for continuous health monitoring of every complete safety instrumented function (SIF). Identifying and predicting problems in the sensors, logic solvers, final elements, and the surrounding process is critical. Sending this information quickly to people who can take corrective action is equally important. Hence, to ensure SIS is effectively successful, the instruments supplied should be IEC 61511 and IEC 61508 certified; also, the supplier organization should have an established management of functional safety.

Advanced safety in offsite (utility/storage/dispatch) area explained through challenges and mitigationsin Table 3.15.

TABLE 3.15

SIS challenges vs. mitigation

Challenges	Mitigation
1. Avoid tanks' overfill incidents:	
Level measurement accurate?	SIL 3 certified 2x or 2 in 1 level measurement to be ensured.
Secondary measurement available?	Single level measurement with guided wave radar or level switch.
Overfill switches working?	Centralized wireless valve position feedback via AMS device manager.
Operator to recognize alarm quickly and can view valve position?	AMS device dashboard to improve visibility of developing overfill to allow preventive action.

2. Minimizing impact of tank leaks:	
How quickly can leak be detected? How much remediation cost if detection is slow?	Use wireless transmitter with fast fuel sensor and AMS to monitor for product around the base and seals of the tank. Also, monitor inventory and alarms in tanks system.
3. Avoiding personal/equipment harm and operation errors:	
How often product needed to be transferred and valve position not known without going to field?	Use smart wireless position monitor. Use alerts directed to maintenance personnel for quick action.
Is it needed to open or close external floating roof drain valve without knowing its current position?	Device status diagnostics and expert guidance from AMS as displayed on a PC in maintenance shop. Plant staff no longer needs to go to field to check.
4. Eliminating loss control monitoring error:	
Is calculation of volume in storage and transfer affected by stratification of temperature in the product or amount of water in bottom?	Use average temperature in calculation as required by standard. Transmitter or sensor to be selected as follows-Multi-spot temperature transmitter & sensor with water cut gauge.-4 channel high density temperature transmitter.-Multiple single point temperature transmitter.
5. Improving pump maintenance efficiency:	
Are the technicians losing valued man-hours in checking vibration?	Automate and centralize pump maintenance by using wireless Vibration transmitter and AMS to detect and report shaft imbalance, misalignment, loose-ness, and machine defects like bearing and gear issues.
6. Avoiding downtime:	
Are failure of field equipments causing shutdown? Are preventive & scheduled maintenance taking long time?	Use AMS supported by smart wireless field device diagnostics to provide predictive warning of issues, and corrective guidance.

Note: AMS stands for advanced maintenance system.

3.6 Adherence to Standards in Electrical Equipment and Engineering

Reduction of hazards is not absolute. There is no absolute safety. Since flammable substances and oxidizers cannot be frequently eliminated with certainty, inhibiting ignition of a potentially explosive atmosphere can eliminate danger at the source.

The objective of selecting an electrical apparatus and means of installation is to reduce the hazard of an electrical apparatus to an acceptable limit. An acceptable level might be designed as selecting protective measures and installation means to ensure that the probability of explosion is not greater due to the presence of electrical apparatus than it would have been had there been no electrical apparatus present.

The most certain method of preventing an explosion is to locate the electrical equipment outside the hazardous (classified) areas whenever possible. In situations where this is not practical, installation techniques and enclosures are available which meet the requirement of locating the electrical equipment in such areas. This method of reducing hazards is based on elimination of one or more of the elements in the fire ignition triangle.

To avoid electrical fire, three principles of fire triangle should be followed, along with the use of protection methods as shown in Table 3.16. The basic point is to ensure that the parts to which a potentially explosive atmosphere has free access do not become hot enough to ignite the explosive mixture.

Operators of hazardous location plants and their personnel should know when explosions are likely to happen and how to prevent them. The manufacturer of electrical equipment and the executing agencies should also be in line to ensure the safe operation of electrical equipment in hazardous locations.

Regulation on hazardous locations containing flammable gas/vapors, dust and so forth have been done in terms of class/division systems formulated by NEC, CEC, OHSA and NFPA and are in place in the industries as standards to follow. These standards allow the design of electrical equipment to eliminate the risk of explosion hazards. These standards enable to design safe and explosion protected electrical equipment and get accredited though testing at designated centres in order to start production. The different area classification with respect to electrical hazards have been established internationally as shown next.

TABLE 3.16

Principles and protection methods

Sl.	Principle	Protection method
1.	Explosive mixture can penetrate the electrical equipment and be ignited but with measures taken it contains the ignition and prevents nearby spreading.	Confine explosion by: •. Explosion-proof enclosure. •. Dust-ignition-proof enclosure. •. Conduit and cable seals
2.	Equipment to be within the enclosure that prevents the ingress of explosive mixture or from coming in contact with the source of ignition.	Isolate the hazard by: •. Pressure & purging •. Oil immersion •. Hermetic sealing •. Encapsulation •. Restricted breathing
3.	Potentially explosive mixtures can penetrate the enclosure but it is be ignited with sparks, though temperature should be within limits.	Limit the energy using: •. Intrinsic safety •. Pneumatics •. Fibre optics

Class definition:

Hazard location classifications as shown in NEC, CEC and NFPA 70 are narrated as below.

Class I locations - flammable vapors and gases are present.

Class II locations - combustible dusts are present.

Class III locations - ignitable fibers or flying are present.

Division definition:

Location classifications are further subdivided into two divisions 1 & 2 as shown in NEC article 500 and CEC section 18. The divisions have been done in terms of hazardous material with respect to flammability shown as follows:

Division	Definition
Division 1	In this case, ignitable concentration of hazards exists under normal operating conditions and / or hazard is caused by frequent equipment failure / maintenance or repair work or frequent equipment failure.
Division 2	In this case, ignitable concentration of hazards may be present but these are handled and processed in closed containers or closed systems from which they can escape only through accidental rupture or breakdown of such systems.

Group definition:

Class I and class II divisions are further subdivided into groups of hazardous materials. The groups define hazardous materials by rating their flammable nature in relation to other standard hazardous materials.

Combustible vapors and gases are divided into four groups. The classification is based on maximum explosion pressure, and maximum safe clearance between parts of a clamped joint in an enclosure as per NEC section 500-5(a)(4).

Class comparison according to zone and class/division are shown in Table 3.17. The group classifications in relation to Class & Division are shown in Table 3.18.

Conclusion from Table 3.18: refineries use group B & D motors only whereas petrochemical plants use group, B, C and D motors. Group A motor is required in neither of these two industries.

TABLE 3.17

Zone vs. Class/Division

Zone	Class/Division
IIC: Acetylene and Hydrogen	A: Acetylene
IIB: Ethylene	B: Hydrogen
IIA: Propane	C: Ethylene
	D: Propane

TABLE 3.18

Group classification in relation to Class and Division

Class	Division	Group	Flammable material	Max. experimental safe gap (MESG)	Min. igniting current ratio (MIC)
Class I	Division 1&2	A	Acetylene	---	---
Class I	Division 1&2	B	Hydrogen, Butadiene, Oxides of Ethylene & Propylene	< = 0.4 mm	< = 0.4
Class I	Division 1&2	C	Ethylene	>4.5 mm	>0.4
			Cyclo-propane Ethyl ether	< = 7.5 mm	< = 0.8
Class I	Division 1&2	D	Propane, Butane, LPG, Natural gas, Acetone, Ammonia, Benzene, Ethanol, Methanol	> 7.5 mm	> 0.8

TABLE 3.19

Permissible temp. vs. temp. class

Permissible surface temperature of electrical equipment, °C	Temperature class
450	T1
300	T2
280	T2A
260	T2B
230	T2C
215	T2D
200	T3
180	T3A
165	T3B
160	T3C
135	T4
120	T4A
100	T5
85	T6

Temperature class definition: Temperature class definitions are defined with respect to the maximum operating temperatures on the surface of the equipment which should not exceed the auto ignition temperature of the substance in the surrounding atmosphere.

Temperature class marking on the equipment means it shall not exceed the ignition temperature of specific gas or vapor to be encountered as specified in NEC section 500-5(d).

A table of permissible surface temperature of electrical equipment versus temperature class is shown in Table 3.19.

Conclusion from Table 3.19: in refineries and petrochemical plants, temperature class of only T5 and in some exceptional cases T6 only are applicable.

Various protection methods as per NFPA, NEC AND CEC:

The principles of protecting the electrical equipment against ignition or explosion are described as follows:

A. **Confine the explosion:**

Enclosures are used a where a hazardous substance can enter the enclosure and get ignited by an electrical spark or hot surface but the explosion is confined within the enclosure.

Refer NEMA standard publication 250, Enclosure for Electrical Equipment, in this regard.

A1. *Explosion-proof enclosure:*

In this case, the enclosure is capable to withstand an explosion of a specified gas or vapor through sparks, or due to an explosion of the gas or vapor within the enclosure, or if the enclosure is located at such a location of temperature where auto ignition would not take place.

Type 7 enclosures are designed to meet explosion-proof requirements for indoor use in locations classified as class I, groups A, B, C or D type. They can also withstand pressure resulting from an internal explosion of specified gas or vapor, containing such an explosion sufficiently that an explosive gas-air mixture existing in the atmosphere surrounding the enclosure will not be ignited. Enclosed devices shall not cause external surfaces to reach temperatures capable of igniting an explosive gas-air mixture in the surrounding atmosphere.

Refer NEC article 100 to get the safety design features of explosion-proof enclosure.

A2. *Dust ignition-proof enclosure:*

Article 502-1 of NEC provides the following definition for dust ignition-proof electrical installations.

Here there shall be no ignitable dust within the enclosure and no spark, heat can enter the enclosure.

Like the above, the enclosure used here is a Type 9 enclosure and is intended for locations classified as class II, group E, F, or G.

Refer NEC article 502-1 to get the safety design features of dust ignition-proof enclosure.

A3. *Conduit and cable seals:*

The NEC guidelines provide criteria for sealing in each conduit run entering an enclosure. These seals are to be installed within 18-inch of an enclosure to avoid propagation of flames and explosive pressure from the interior of the enclosure into conduit system.

B. **Containing energy in case of explosion/ignition:**

There are various defined safety features, as discussed below:

B1. *Intrinsic safety:*

It should be considered in integration with an electric circuit. Here, equipment and wiring shall not release electrical or thermal energy to cause ignition of a hazardous material in the atmosphere. Intrinsically safe systems contain the energy using an energy-controlling apparatus known as an intrinsically safe barrier in situations of wiring or component failure to prevent ignition.

B2. *Pneumatic safety:*

This system uses air in a control circuit instead of using a power cable. Hence, there shall be no ignition source even if hazardous substances are found in the atmosphere.

B3. *Fibre optic safety:*

Here, optical energy is contained in a control circuit instead of in a power cable. Hence, there shall be no ignition source even if hazardous substances are found in the atmosphere.

C. **Hazard isolation safety:**

Here, the system either prevents or delays the diffusion of a hazardous substance into an enclosure. There are various kinds of designs in this category, as detailed below:

C1. *Pressurization and purging:*

Here, the enclosure is kept under inert gas pressure with continuous purging facility to avoid the ingress of hazardous substances.

C2. *Oil immersion safety:*

In this system, the electrical equipment is immersed in inert oil that quenches any spark or flame and control surface temperature to a safe level. The most common example is a power transformer.

C3. *Hermetic sealing:*

In this system, devices of the enclosure are sealed, soldering, brazing, welding, or fusing the glass to metal to avoid ingress of external hazardous substances.

C4. *Encapsulation safety:*

In this system, devices of the enclosure are encased by a moulding material to prevent any spark from coming in contact with hazardous substances.

C5. *Restricted breathing safety:*

Here gasket type sealing is used in situation where inflammable gas in the surrounding can't stay for a longer duration for causing ignition.

Comparison of various protection methods:

- Intrinsic safety is highly reliable system but careful planning and design is required for installation in a hazardous atmosphere like petroleum refinery.
- Explosion-proof system is most reliable in very hazardous atmosphere like petroleum refinery but it is most costly.
- Pressurized purging system is used in the petroleum refinery to ensure special safety like alarm and tripping of the equipment on low pressure of inert gas.
- Oil immersion system is a very important method and applied in electric transformers installation.
- Hermetic sealing is a low-cost system and operates at reduced current.
- Encapsulation is also a low-cost system and components are reusable.
- Restricted breathing is also a low-cost system but has gasket failure potential.
- Pneumatic system is also a low-cost system, easy to install but control response time is slow.
- Fiber optic system is ideal for a dust-free atmosphere but has limitations with respect to distance the light must traverse for the control circuit to work.

Comparison of class, division, and zone system:

Class/Division standard is not the only standard that exists for hazardous locations applications. There is another standard known as the Zone standard that is used mainly outside North America.

Class/Division methods are used in North America with requirements set by NEC/CEC, although the Zone system has wider use throughout the world in the chemical and petrochemical industries. The Zone methods cover more complicated situations unlike class/division methods with respect to handling a particular application because zone methods take care of the type of protection that equipment requires.

NEC and CEC govern North American standards; in Europe, it is European (EN) standards; other countries work with standards based on international standards governed by IEC or accept products based on standards like NEC, CEC or EN.

Comparison of IEC/NEC/CEC Zone standards with NEC/CEC Class/ Division standards & area classification are shown in Table 3.20.

As per NEC article 505-10(b)(1), a Division classified product may be installed in a Zone classified location but the revere is not true. Typically a Zone-classified product provides protection utilizing a protection method not specified in Class/Division scheme.

TABLE 3.20

Comparison between Zone standards and Division standards

Zone 0	Zone 1	Zone 2
Ignitable concentration of flammable gases, vapor or liquid are present continuously.	Ignitable concentration of flammable gas, vapor or liquid:-present under normal operating conditions.-may exist frequently due to repair maintenance, operation or leakages.	Ignitable concentration flammable gas, vapor or liquid:-present under normal operating conditions.-occur only for short duration. -hazardous only in case of an accident or some abnormal conditions.
←--------------------------------Division 1--------------------------→		←--------Division 2-------→
Ignitable concentration of flammable gas, vapor, or liquid:-are likely to exist under normal operating conditions.-exist frequently because of repair maintenance or frequent equipment failure.		Ignitable concentration of gas, vapor, or liquid:-are not likely to exist in normal operating conditions.-normally in closed enclosure where the hazard can only escape through an accidental rupture or breakdown of such containers or abnormal operating conditions.

Summery comparison of IEC, NEC and CEC methods is shown in Table 3.21. Summery comparison of NEC/CEC & IEC grouping are shown in Table 3.22. Comparison of Zones and Division with respect to temperature class are shown in Table 3.23.

Conclusion: In refineries and petrochemical plants, motors' temperature class don't change with variation in Zones or Divisions as seen from the above table because temperature remains ambient or 100^0C in exceptional cases, i.e., temperature class is restricted to T5 or T6 only.

Standards followed with respect to electrical hazards:

- IEC 60079, ANSI/ ISA-91.00.01-2001, ANSI/ ISA-TR99.00.01-2007, ANSI/ ISA-18.2-2007, ANSI/ ISA-61010-1 (82.02.01)-2004, ISA-TR84.00.07-2010, ISA-TR98.00.02-2007: Applicable both for electrical and instrumentation.
- NEC/CEC and Zone standards as mentioned above.
- NFPA 70, 70B, 70E, 77, 78 & 79.
- OISD STD 110, 113, 149, 173, 180-Electrical Safety related in India.

TABLE 3.21

Comparison between NEC/CEC method and IEC method.

Presence of hazards	NEC/CEC method	IEC method
Continuously	Division 1	Zone 0
Intermittently		Zone 1
Under abnormal conditions	Division 2	Zone 2

TABLE 3.22

Comparison between NEC/CEC Grouping and IEC Grouping.

Typical hazard	NEC/CEC grouping	IEC grouping
Acetylene	A	IIC
Hydrogen, Butadiene, Ethylene, etc.	B	IIC
CO_2, SO_2, etc.	C	IIB
Gasoline, Ammonia, Ethanol, Propane, etc.	D	IIA

TABLE 3.23

Comparison among Zone, Division and Temperature.

Zone 0, 1 & 2	Division 1 & 2	Maximum temperature, °C
T1	T1	450
T2	T2	300
	T2A	280
	T2B	260
	T2C	230
	T2D	215
T3	T3	200
	T3A	180
	T3B	165
	T3C	160
T4	T4	135
	T4A	120
T5	T5	100
T6	T6	85

3.7 Plant Operation

Operating and pocket manual:

Operating manuals should be available to operating crew for safe operation of each process unit and offsite. The operating manuals should be approved by authorized dignitaries of the organizations. Also, to facilitate easy and quick reference, people directly engaged in the operation should be provided with handy pocket manuals.

Quality control:

Operating crew should work in tandem with quality control personnel, as described in operating manuals. Operation in charge should ensure that all test samples are analyzed in the right frequencies as already decided and act accordingly as per manual and operating instructions. They should maintain total records of all data and events.

Checklist for startup, shutdown, known emergency handling:

Operation in charge should ensure preparation and approval of these documents so that during respective activities these can be referred to instead of re-inventing the wheel.

Bypassing any safety interlock:

A record register should be in place for bypassing safety interlock with approval from a competent authority and re-instatement of the same at the earliest with everyday review on the same by the operation in charge.

Adequacy of communication, visibility & illumination:

PA system, walkie-talkie should be available for the operating crew. Also, CCTV should be available to ensure continuous visibility of the vulnerable operation areas.

Work permit system:

Ownership of any operation unit lies with operation in charge; hence, any maintenance or any service job to be executed inside the unit must have a duly approved work permit taken by the executor. Hence, a work permit system incorporating all checklist points from a safety point of view should be developed and strictly followed.

Critical equipments monitoring:

A checklist of all critical parameters should be prepared for continuous monitoring of the same though operation log sheets developed by the operation management should be religiously filled in at regular intervals, as decided by a competent authority.

Operating instruction:

Operation in charge should maintain a record of providing operating instructions to the operation crew.

MSDS:

All MSDSs should be available as provided by the designer and by the quality control department.

Traceability of documents:

All documents as discussed above should be traceable for easy access, as and when necessary.

Safety tour of units at combined levels:

A competent authority should establish a system of safety tours by different designated persons at specified intervals to ensure third-party vigilance on the operation.

Management safety review:

There should be system of first incident reporting to be ensured by the operating stuff directly engaged in the operation; a competent authority should subsequently constitute an investigation committee for root cause analysis and issuing of the report. This not only ensures discipline but also establishes a system of lesson learning to avoid the occurrence being repeated.

PPE compliance:

Personal protective equipment (PPE) should be used by any person entering the inside operation area. A competent authority should also declare the hard hat areas and ensure display of the notification.

3.8 Plant Maintenance and Inspection

- OISD standards for rotary and stationary equipments, as mentioned in section 3.5 above, should be followed.
- Make small checklists for daily monitoring of critical equipments.
- Ensure adherence to work permit system, as discussed above.
- Ensure PPE compliance, as discussed above.
- Refer material safety datasheet (MSDS) to handle chemicals and take precautions accordingly. For e.g., for toxic HC, use portable gas detector to enter into unit; for heavier HC, use gas detectors at ground level; for lighter HC, take special precaution, e.g., for H2 leak, never try on line rectification by tightening; except with bubble leak, use brass hammer and continuous water pouring on leaky flanges.
- For inspection activities in various areas, follow standard procedures; examples of checklists for various civil engineering inspection activities are enumerated below, as specified in OISD STD 170.
- *Checklist for inspection of buildings:*

	Equipment no:	Name:	
	Date:	Location:	
Sl.no.	Components	OK/Not OK	Remarks
	Cracks/settlement of foundation wall		☐

Floor	☐
Roof	☐
Structure	☐
Doors and windows	☐
Paintings	☐
Fittings and fixtures of toilet, WC	☐
Other fittings	☐
Choking of drains, septic tanks, etc.	☐
Water leakage/seepage	☐
Plinth protection around the buildings	☐
Minor electrical/wiring in the building	☐
Steel structure	☐
Insulation/partition, false flooring, ceiling	☐
Spark less flooring	☐
Construction and expansion joints	☐
Any other	☐

Roof

1	Growth of vegetation	☐
2	Blockage of gutter and rain water pipes	☐
3	Damage of water/weatherproof treatment	☐
4	Damage of expansion joints	☐
5	Damage of slabs, expansion joints due to vibration/earthquake	☐
6	Debris/discarded equipments	☐
7	Overhead water tank overflow	☐
8	Water logging	☐
9	Crack on AC/CI/GI pipes	☐
10	Crack on roofing sheet	☐
11	Cleanliness of transparent sheet	☐
12	Loosening/lack of sheet bolts/washers	☐
13	Crack on RCC slabs and consequent leakage	☐
14	Any other	☐

- *Checklist for inspection of equipment foundation:*

Equipment no:	Name:
Date:	Location:

Any visible crack in foundation	☐
Excessive vibration during	☐
Running of equipment	☐

Calcinations or deterioration of concrete	☐
Physical damage of foundation	☐
Loosening or corrosion of foundation bolts	☐
Exposure and corrosion of reinforcement	☐
Staining of concrete	☐
Any spillage of product on foundation	☐
Stagnation of any oil/water/liquid near	☐
the foundation	☐
Any unusual or abnormal settling such as:	
A sloping floor	☐
Cracks in floors and walls	☐
Displacement of some parts	☐
Piping out position	☐
Piping under strain	☐
Broken structural bolts	☐
Top levels of saddles/concrete	☐
Columns of foundations with	☐
respect to benchmark	☐
Any other	☐

- *Checklist for inspection of fireproofing:*

Equipment no:	Name:
Date:	Location:

Any visible crack in the coatings	☐
Selectively remove small section	☐
of fireproofing and check condition	☐
of the face of substrate	☐
Check condition of reinforcing wire	☐
Check the loss of fireproofing	☐
materials as a result of mechanical	☐
abuse	☐
Check the bulging out of fireproofing	☐
materials	☐
Check for any discoloration	☐
Whether any rust strain observed	☐
Check the abnormal sound against	☐
tapping with a light hammer	☐
Check the condition of joint sealant	☐
Check the condition of weather coating	☐
Any other	☐

- *Checklist for inspection pipe racks and tracks:*

Equipment no:	Name:
Date:	Location:

Vegetation growth	☐
Soil contract	☐
Water logging	☐
Settlement	☐
Stagnation corrosion	☐
Pitting on structural members	☐
Perforations	☐
Misalignment of pipes	☐
Buckling of column/support of pipe racks	☐
Deflection of beam of pipe racks	☐
Cracks on slab near column due	☐
to punching	☐
Displacement of anchor/supports	☐
Effect of addition/alterations of pipe racks	☐
Vibration of supporting members	☐
Damage to fireproofing	☐
Honeycombing of RCC members	☐
Spalling, discoloration of concrete	☐
Any other	☐

- *Checklist for inspection of tank foundation and dyke walls:*

Equipment no:	Name:
Date:	Location:

Tank settlement:	☐
Adequacy of drainage system:	☐
Grass/shrub growth on tank pad/apron erosion of tank pad/apron:	☐
Checking of chemical analysis of concrete for ruling out alkali aggregate:	☐
Reaction induced cracks:	☐
Condition of joint between tank bottoms and foundations:	☐
Maintenance and upkeep of tank farm area, including pathways:	☐
Vegetation growth of tank farm area:	☐
Spillage of any tank contents in tank foundation which may damage or erode tank foundation:	☐
Any other:	☐

- *Dyke Walls*

Checking of grass/shrubs on dyke walls ☐
Erosion and loss of heights ☐

Condition of joints in case of masonry or concrete dyke walls ☐
Drainage system in and around dyke walls ☐
Cracks on masonry/RCC walls ☐
Erosion of soils around foundation of dyke walls ☐
Any other ☐

3.9 Firefighting: Operation, Standards and Practices

Even after all possible safe design and practices followed in the industry, there can still be a chance of fire breaking out inside the process plants, offsite or in any office building of the industry. Hence, adequate firefighting facilities should be available to tackle emergencies such as fire if it happens.

To establish the right facilities for firefighting, it is necessary to understand the classification of fires and accordingly establish firefighting facilities and train operating and maintenance staffs to operate and maintain the facilities effectively.

3.9.1 Classification of Fire and Application of Fire Extinguisher Media

There are four types of fire as follows:

i. **Class A fire**
Fire catching up in combustible materials like paper, wood, etc. This is called solid fire and water is the correct extinguishing medium.

ii. **Class B fire**
Fire catching up in flammable vapor-air mixtures, which may accumulate on the surface of flammable liquids such as petroleum fuel, paints and varnishes. This is called liquid fire. Using water as an extinguishing medium may spread the fire. Hence, dry chemical powder (DCP), foam pourer, CO2, and halogenated hydrocarbons are found to be very effective as extinguishing medias.

iii. **Class C fire**
Fire catching up in gaseous material like vapor clouds of hydrocarbon or any volatile organic substances. This is called a gas fire. DCP, CO2 and vaporizing liquids are used as extinguishing media in case of this type of fire.

iv. **Class D fire**
There are some combustible metals also which catch fire; for example, magnesium, titanium, zirconium, lithium, sodium and even a thin film of iron in some cases. This is called a metal fire. Tertiary eutechtic chloride (TEC) is very suitable in this case. However, water is also used as a general firefighting medium.

Electrical fires also fall into this category; however, to fight electrical fires, DCP & CO2 are mostly used and sometimes halons also are used.

Principles in application of Fire Extinguishers

We know that the principle of firefighting is cooling, smothering; i.e., blanketing the fire from atmospheric oxygen, and starvation; i.e., cutting of the fuel source whatever may be the class of fire mentioned above. But to meet these principles, various extinguishing media are found effective in respective cases, as mentioned earlier.

The cheapest and most effective extinguishing medium is water, which achieves the best cooling effect unlike other media while at the same time it achieves the purpose of smothering also.

But in some special cases like electrical fire particularly and in a flammable liquid pool, water is not used in order to prevent electrical short circuits and splashing of inflammable liquid, thereby spreading fire. Hence, to fight electrical fire, CO2 or DCP are used as extinguishing media; but both these two are costly and so these are used in relatively small fires or used before the fire spreads. When the fire spreads, using water is the only available source of extinguishing; in such a situation, ensure that electrical power supply is cut off first before spraying water.

Also, to extinguish fire from a large pool of liquid, foam pourer should be used instead of direct spraying of water due to reasons mentioned earlier. Here, to get the best performance on starvation, i.e., cutting fuel from fire, foam pourer provides a very good result, i.e., though it is sprayed in water mix, a layer of foam remains on the fuel surface while much of the water layer may evaporate out.

Types of foam pour used in the industries

The foam concentrate contains suitable stabilizer and sufficient preservative to prevent decomposition due to micro-biological attacks during storage. The foam cannot be stored for a long period and generally, its shelf life is about three years. There are four types of foam chemical used in the industries, as described below:

- *Protein foaming agent (P):*

  It is comprised of high molecular weight natural proteinaceious polymers derived from chemical digestion and hydrolysis of natural protein solids. The polymers give elasticity, mechanical strength and water retention capacity to the foam. The dissolved polyvalent metallic salts in concentrate aid in foam bubble strengthening. It is available for proportioning of about 3 to 6% by volume in water.

- *Fluor protein foaming agent (FP):*

  This is similar in composition to above with respect to protein foam concentrate, but additionally it contains a fluorinated surface active agent that contributes a fuel shedding property to the foam generated. This enables its use for sub surface injection for tank firefighting where foam may plug into the fuel.

- *Aqueous film forming foaming agent (AFFF):*
  This is composed of synthetically produced materials that form air foams similar to protein foam. But these foaming agents are capable of forming water solution film over the surface not fully covered with foam, thus serving better than FP mentioned above. In petroleum industries, it is very commonly used.
- *Alcohol type foaming agent:*
  Air foams generated from ordinary agents rapidly break down, resulting in loss of effectiveness in firefighting in cases where fuels are water soluble or water miscible or have polar solvent; i.e., various flammable chemicals like alcohol, MEK, MIBK, amines, hydrate, etc.
  To fight fire in such chemicals, alcohol type or polar concentrates are used.
  Foam concentrates as mentioned above are proportioned to a final concentration of about 3 to 6% by volume using fresh or sea water.
  The qualities of the foam concentrates are classified in terms of their expansion ratio which is very important, as shown below:
  - Low expansion foam has expansion ratio of 8:1
  - Medium expansion foam has expansion ratio of 450:1
  - High expansion foam has expansion ratio of 1000:1

The ratio is maintained by the proper design of foam-making equipment which ensures the optimum ratio of prescribed water pressure.

Note: the success of using foam depends upon its consumption which should not be more than 1 GPM/Sq. Ft.

Method of operation of extinguishers/Fire water hoses

CO_2 extinguisher

Extinguisher is first to be held at an upright position, then the locking ring pin is pulled, and the discharge nozzle operated in one hand (or the valve is unscrewed, if designed like that) and holding the extinguisher cylinder in the other hand. The discharge nozzle/horn is connected by a high pressure flexible hose to the cylinder.

Direct the jet of released CO_2 at the fire source and sweep the horn frequently around the source. The horn direction should be downward to the spo-t of fire to facilitate better dispersion of CO_2 throughout the fire source instead of short circuiting the source. Minimum discharge time for portable extinguisher varies from eight to thirty seconds and maximum range of spray varies from 1 to 2.5 meters.

CO_2 extinguisher is best recommended to douse the fire from an electric short circuit in the zone of an electric circuit; however, the power supply switch should be cut off the first if it is approachable. For large fires out of the electric short circuit, it may not be possible to douse fire by the extinguisher. In that case water or foam-water jet spray at a high rate from a fire tender would be necessary; before using water as a firefighting medium, the main power supply must be cut off.

DCP (Dry chemical powder) extinguisher

In a DCP extinguisher, there is an inner cartridge inside the cylinder which contains a mixture of sodium bicarbonate, potassium bicarbonate and potassium chloride.

To operate, the extinguisher should be held at upright position holding it in one hand. Remove the safety clip and strike the knock to actuate the piercing mechanism, which in turn breaks the sealing disc of the CO_2 cartridge while holding the discharge nozzle in one hand. The discharge is a white fume of mixture of CO_2 and residual solids from the chemical after release of CO_2. Direct the spray to the base of the fire, move the nozzle around it keeping the nozzle mouth downward to the base of the fire to facilitate dispersion of CO_2 through the fire source instead of bypassing the base. It is also applicable for dousing a small fire or extinguishing the fire before it escalates.

Operating fire water hose

Fire water hose to be taken out from the hose box located near the fire water hydrant (stationary fire water supply point fitted with isolation valve and nozzle). Unroll the hose and the male joint fitted in one end of the hose should be fixed with the female joint fitted in the hydrant. Then approach towards the fire as much as it is safe to proceed, and then with both hands, tightly hold the nozzle located at the other end of the hose keeping the nozzle underarm and pressing the same with arm pressure. In this condition, the other operator at the fire hydrant end should open the valve of the hydrant when the operator at the nozzle end as mentioned above can spray water to the fire to douse the same. The tight holding procedure as above is needed as hydrant water pressure will be minimum 7 bar; if the procedure is not followed, the spray nozzle may hit the leg , hand or parts of the body of the operator who is spraying the water to fire.

3.9.2 Firefighting Facilities and Their Upkeep in the Petroleum Industry

While various extinguishers are to be kept available in each process plant, control rooms, office rooms and buildings in designated locations as per industry safety guidelines and stipulations, the following other facilities have to be established inside at the boundary limit of the industry to manage any small or large fire, as discussed below:

i. Fire water network throughout the periphery of all process plants and offsite facilities along with connecting roadside. At a regular interval of distance, there should be a fire hydrant nozzle and there will be a hose box placed adjacent to the nozzle so that when necessary, that hose can be connected to the nozzle to deliver water spray to the location of fire. In the fire hydrant nozzles surrounding the process plants where there is elevated equipment, fire monitors of designated capacity/specification are fitted to the nozzles to support the fire equipment for water spraying; for e.g., medium volume long range (MVLR) or high volume long range (HVLR) monitors are in use.

ii. Equipment handling/storing critical hazardous hydrocarbon/chemical should be provided with a water sprinkler in line with industry safety regulations to ensure sufficient cooling of the equipment in case of fire in the vicinity. These sprinkler systems should be designed as per stipulated guidelines with respect to adequacy of flow, covering full surface and with auto start facility on thermal sensors activation and with remote start/stop facility.

iii. Two fire water ponds of adequate capacity should be provided so that during de-silting of one pond, the other can be available.

iv. A fire water pump house with adequate capacity and number of pumps should be provided connected to the fire water network as mentioned above. Details are as below:

- Two small pumps, generally called Jockey pumps, of which one should run and the other should be as standby. Even one should not run all the time — in a petroleum refinery, fire water network pressure is generally kept at 7 bar; when this header pressure comes down below this target pressure, the jockey pump is auto started to maintain the header pressure. So, if the jockey is not auto stopped or frequently gets auto started, the concerned operating crew should find the water leak source and rectify it accordingly.
- In addition to this jockey pump, two or more bigger pumps also to be provided should be started in auto mode, which occurs during fires when the continuous jockey pump running also decreases the header pressure below 7 bar due to a large quantity of water consumed in firefighting. When the bigger pump starts, it is generally set to enhance the header pressure at a higher level, say at 11 to 12 bar in the case of a petroleum refinery or petrochemical plant; if one bigger pump cannot maintain that desired header pressure, the second bigger pump should take auto start to maintain header pressure to enable plant people to fight the fire properly.
- All these pumps generally are motor driven; but, against two motor driven bigger pumps, two standby diesel generator driven pumps should be provided so that, in case of non availability of electric power, fire water pumps can be operated for firefighting.

v. The industry should have an adequate number of fire water tenders which should reach out to the place of fire when called for. In a refinery or petrochemical plant, generally seven or eight such fire water vehicles are provided; whereas for petroleum depots and terminals, two and four vehicles respectively are found to be adequate.

vi. The firefighting control room should be established to operate and manage fighting inside the industry boundary, as well as also to assist neighboring areas in case of fire outside the boundary. The fire water tenders, as mentioned above, are kept adjacent to fire control room under the jurisdiction of the fire control room. The fire control room is manned by an operating crew which monitors the fire water header pressure, any alarm signal coming from the plant, receives any emergency calls from any location and acts.

vii. Regarding alarms, mentioned above, it is worth pointing out that in process plants and in some designated locations alongside the fire water network, there are some ananciator pill boxes kept; in each pill box, there is a switch pressed by a cover glass which is broken by any person observing fire nearby; on breaking the glass cover, the switch spindle comes out and consequently, a hooter sound started buzzing in the vicinity as well as to process control rooms, office buildings and fire control rooms to help people act accordingly.

The plant and offsite areas covering hazardous storage facilities are also provided with gas explosimeter which are generally placed at the ground level at designed locations, as evaluated by a competent expert organization; these meters send hooters to plants and fire control rooms to help them initiate response.

Special type of alarms are provided at the top of floating roof storage tanks which act in auto mode on thermal sensor activation in case of fire, as well as explosimeters to detect vapor cloud.

viii. All control room and offices should be provided with a first aid box.

ix. Plant personnel should wear safety shoes, hand gloves and a helmet while entering the plant area and use safety goggles while working in flame areas, such as when operating a furnace. They should also ensure welders use safety goggles if any such job is allowed with the permission of a competent authority.

x. The industry should establish a firefighting organogram to take care of everything mentioned above. They should also keep control of inventory of all types of fire extinguishers, foam liquids, firefighting aprons, hand gloves, safety shoes, safety helmet, safety goggles, fire blankets, portable breathing apparati, rescue ropes and spider nets, first aid materials, etc.

xi. The firefighting department should monitor each extinguisher at all locations at a regular interval to ensure refilling of the same and to reject the same if a defect is found as well as check whether the expiry date of the extinguisher is reached. They should also check the first aid box at regular intervals at all locations to refill the same whenever required.

3.9.3 Evaluation of Fire Water Flow Rate Requirement in Refinery

The design basis has been revised by Oil India Safety Directorate (OISD) for India and referred in standard, OISD-116 in view of large fire occurred in Jaipur oil depot on 29th October, 2009. However, for India as well as for other countries also, all practices and standards are generally found as corollaries of respective standards of the international organization, NFPA (National Fire protection Association).

The calculation of fire water requirements as per revised OISD standards are shown as below:

The fire water system is designed to meet the fire water requirement to fight two major fires simultaneously or a single fire for the largest floating roof tank's roof sinking case, whichever requires the largest water demand, as per clause no. 5.1 of OISD-116. The calculation to evaluate fire water requirement is given as an example as follows:

The calculation is done tank farm wise with figures of each category designated under section A, B1, B2, B3, C, D & E as follows and comparisons done accordingly to derive the required water flow. Finally, equation F gives the desired capacity of flow.

Basis considered: Fire in Floating roof tank.

A. **Crude Oil Tank:**

Case-I: Fire water calculation for full surface fire on largest floating roof tank (roof sinking case):

Note: The tanks are placed inside the tank dyke area; (dyke is an elevated enclosure of about 0.6 meter height surrounding the tank (s) meant for holding the content of largest capacity tank in eventuality of oil spillage/leak from the tank).

a. Data:

Total storage capacity in one tank dyke area:	1,00,000 M3
No. of tanks in one dyke area:	2
Capacity of each tank:	50,000 M3
Diameter of each tank:	70 M

Height of each tank:	14.4 M
Type of tank:	Floating roof

b. Cooling water requirement:
Cooling water rate of tank shell area for tank on fire= 3 LPM/M2 (LPM- Liters per minute)

Cooling water required = 3.142 × 70 × 14.4 × 3
= 9501.4 LPM
= 570 M3/Hr.

The 2nd tank is located within dyke area at a distance more than 30 M from the tank shell.
Then cooling water required @1 LPM/M2 of tank shell area

=3.142 × 70 × 14.4 × 1
=3167 LPM
=190 M3/Hr.

Then total cooling water rate = 570 + 190 M3/Hr.
= 760 M3/Hr.

c. Water requirement in foam application:
Foam application rate= 8.1 LPM (as per NFPA-11)

Foam solution requirement = (3.142 × 70 × 70)/4 × 8.1
= 31176.5 LPM
= 1870 M3/Hr.

Water required for foam solution = 97% × 1870 M3/Hr.
= 1815 M3/Hr.

d. Fire water for supplementary hose stream based on 4 hydrant streams+2 nos. HVLR monitor:

4 × 36 M3/Hr + 2 × 228 M3/Hr
= 600 M3/Hr.

e. Total water required for roof sink case:

Tank cooling:	760 M3/Hr.
Foam application:	1815 M3/Hr.
Supplementary stream:	600 M3/Hr.
Total:	3175 M3/Hr.

Case-II: fire water flow rate for floating roof tank protection:

a. Data:

Total storage capacity in one dyke area:	1,00,000 M3
No. of tanks:	2
Capacity of each tank:	50,000 M3/Hr.
Diameter of each tank:	70 M
Height of each tank:	14.4 M

b. Cooling water requirement:
Water rate for tank shell area for tank on fire= 3 LPM/M2

$$\text{Cooling water requirement} = 3.142 \times 70 \times 14.4 \times 3$$
$$= 9504.4 \text{ LPM}$$
$$= 570 \text{ M3/Hr.}$$

The 2nd tank is located within dyke at a distance of 30 Meter from the tank shell. Then , water requirment @1 LPM/ 2 of tank shell area becomes: $3.14 \times 70 \times 14.4 \times 1$

$$= 3167 \text{ LPM}$$
$$= 190\, M3/\text{HR.}$$

Hence, Total cooling water flow

$$= 570 + 190\, M3/\text{HR.} = 760\, M3/\text{HR.}$$

c. Foam water requirement for rim seal area (at tank roof): (Water flow required for applying foam on largest tank surface)

Diameter of the tank=	70 M
Distance of foam dam from shell=	0.8 M
Diameter of roof up to foam dam=70-(2x0.8)=	68.4

$$\text{The rim seal area} = (3.142/4) \times (70^2 - 68.4^2)$$
$$= 173.9 \text{ M2}$$

$$\text{Foam solution rate @12 LPM/M2} = 12 \times 173.9 = 2086.8 \text{ LPM}$$
$$\text{Water required for 3\% foam solution} = 0.97 \times 2086.8 \text{ LPM}$$
$$= 2024.2 \text{ LPM}$$
$$= 122 \text{ M3/Hr.}$$

d. Fire water for supplementary hose stream based on 4 hydrant stream + 2 HVLR monitors:

$$4 \times 36 \text{ M3/hr} + 2 \times 228 \text{ M3/Hr.} = 600 \text{ M3/hr.}$$

e. Total water required:

Tank cooling: 760 M3/hr.
Foam application: 122 M3/hr.
Supplementary stream: 600 M3/hr. (A)
Total: 1480 M3/Hr.

B. Other tank farms inside the refinery:
Similarly, calculations to be done for other tank farm and say, following have been found:

For Naphtha + Gasoline Tankfarm – 1475 M3/Hr …… (B1)

For Kerosene, ATF & other Naphtha Tankfarm – 2075 M3/hr … (B2)

For Diesel Tankfarm – 1290 M3/Hr …… (B3)

Note: Here, crude oil, naphtha and gasoline are class C petroleum products and kerosene/ATF & diesel are class B petroleum products. When all class A & B products are considered, other class C products which are less flammable compared to the other two, and where if the total volume is less than the total volume of class A & B products —which is natural in a petroleum refinery—then the firefighting water calculation need not be done for class C products.

C. **LPG sphere:**

a. Data:

No. of spheres:3 (2 in operation).	
Capacity:	2573 M3 each
Diameter:	17 M

b. Cooling water requirement:

Water application rate = 10 LPM/M2

Water required = $3.142 \times 17^2 \times 10 \times 2$

= 18160.8 LPM

= 1090 M3/hr.

c. Fire water for supplementary hose stream based on 4 hydrant streams + 2 HVLR monitors:

Water required = 4×36 M3/hr + 2×228 M3/hr. =600 M3/Hr.

d. Total water required:

Sphere cooling: 1090 M3/hr.

Supplementary stream: 600 M3/Hr. (C)

Total: 1690 M3/hr ……

D. **LPG bullets:**

a. Data:

No. of bullet:1	
Size:	9 M (TTL); TTL means tail to tail distance.
	3 M (ID); ID means internal diameter.

b. Cooling water requirement:

Water application rate = 10 LPM/M2

Water required = $3.142 \times 3^2 \times 10 + 3.142 \times 3 \times (9 - 3) \times 10$

= 848.3 LPm

= 51 M3/Hr.

e. Fire water for supplementary hose stream based on 4 hydrant streams + 2 HVLR monitors:

4X36 M3/hr + 2 × 228 M3/hr = 600 M3/hr.

f. Total water required:

Bullet cooling:	51 M3/hr.
Supplementary stream:	600 M3/Hr.
Total:	650 M3/Hr ……

(D)

E. **Process units protection:**
For process unit protection in case of fire, the largest and most severe hazardous process unit should be considered in the calculation. Also, water is to be applied using fixed water monitors and hose lines. Three following alternatives are considered for fire water rate.
Alternative-I

Hydrocracker + adjacent H2 unit area: $200 \times 80 + 74 \times 80 = 21920$ M2

Water application rate: 1 LPM/M2

Water required = 21920 × 1 LPM

= 21920 LPM

= 1315M3/Hr.

Fire water for supplementary hose stream based on 4 hydrant streams + 2 HVLR monitors:

4 × 36 M3/hr + 2 × 228 M3/hr = 600 M3/hr.

Total water required: 1315 + 600= 1915 M3/Hr.

Alternative-II
Consider 10Mx10M portion of the process unit area on fire. Provide water cover over an area of 30Mx30M at a rate of 10 LPM/M2.

Water required = $30 \times 30 \times 10$ LPM

= 9000 LPM

= 540 M3/Hr.

Fire water for supplementary hose stream based on 4 hydrant streams + 2 HVLR monitors:

4 × 36 M3/hr + 2 × 228 M3/hr = 600 M3/hr.

Total water required: 540 + 600= 1140 M3/Hr.

Alternative-III
Water required for portion of unit area provided with fixed spray system (Extreme hazardous area).

Area assumed = 1000 M2
Water rate = 10 LPM/M2
Cooling water required = 1000 × 10 LPM
= 10000 LPM
= 600M3/Hr.

Fire water for supplementary hose stream based on 4 hydrant streams + 2 HVLR monitors:

4X36 M3/hr + 2 × 228 M3/hr = 600 M3/hr.

Total water required: 600 + 600= 1200 M3/Hr.

Water flow considering maximum of 3 alternatives = 1915 M3/Hr... (E)

F. **Overall design flow rate:**
Fire water flow rate is based on fighting two major fires simultaneously anywhere in the refinery complex. Major hazardous areas have been considered for firefighting and fire water demand for each area has been calculated, as summarized as below:

Crude tank farm	=1480 M3/Hr. ref. Case-II of A
Naphtha/Gasoline tank farm	=1475 M3/Hr....ref. B1
Kerosene/ATF & other Naphtha tank farm	=2075 M3/Hr... ref. B2
Diesel tank farm	=1290 M3/Hr.....ref. B3
LPG sphere	=1690 M3/Hr.....ref. C
LPG bullet	=650 M3/Hr.......ref. D
Unit protection	=1915 M3/Hr.....ref. E

For fighting the above two major fires simultaneously, the design flow rate should be the sum of the two highest flow rates from the above, i.e.,

$$\text{Design flow rate} = 2075 + 1915 \text{ M3/hr.} \quad \text{(F)}$$
$$= 3900 \text{ M3/hr} \ldots \ldots$$

Say 4000 M3/Hr.
For full surface fires of the largest floating roof tank (Roof sinking case): Total fire water flow rate required has been found to be 3175 M3/hr, as shown in case II of sl. A above.
Hence, design flow rate should be the highest of the two, i.e., 4000 M3/hr. So, fire water network capacity should be designed to take this flow.

G. Fire water storage:
The effective pumping capacity of fire water reservoir shall be minimum 4 hours aggregate working capacity of pumps. The reservoir shall have two compartments to facilitate cleaning and repair. So, the reservoir capacity considering the requirement of 4000 M3/Hr will be: (4000x4) M3/Hr= 16000 M3/Hr. Capacity of each compartment shall be 8000 M3 at least.

H. **Fire water pump:**
As per OISD standard-116, clause 5.5.3 (ii), when total numbers of working pumps are more than two, 50% capacity standby pumps shall be provided. Also, as per standard, all main fire water pumps shall be of identical capacity and characteristics.

3.9.4 Training and Development

Fire and safety departments, having their responsibility on this subject, should ensure the following to demonstrate good records on training and development of all employees and stakeholders as below:

- The department should have a training classroom with amenities of training needs and an outdoor field for conducting in-field training on firefighting.
- The department should keep adequate inventory of the CO2 cylinders and DCP cylinders for consumption in the training through planning and keeping of records.
- There should be an induction firefighting training program for all newly joined employees, prioritizing employees joining in the operation first, and next employees working outside battery areas; i.e., in the administrative offices.
- There should be a training calendar for all employees engaged in the operation for retraining or training in new skillsets in firefighting, as approved by the head of the fire & safety department and reviewed by a competent authority to ensure availability of the employees for the same.

- All training/retraining modules should be prepared by the fire and safety department.
- The department also should carry out an annual safety awareness survey through a questionnaire for all employees to evaluate average employee awareness on safety; This in turn helps in determining the human error prediction index in tackling or preventing unsafe situations in that industry.
- The department should promote a safety incident reporting system by the employees such as near miss reporting, incident FIR report and ensuring a permit system for carrying out any job by the service department.
- The department also should introduce a performance scoring system on lead and lag indicators of safety performance.

3.10 Emergency Preparedness Program

A sudden unexpected occurrence demanding immediate action tells us what emergency means.

Emergency is a situation occurring suddenly or progressively and unexpectedly which may cause life, environment and property to be in danger. For fighting such emergencies, some protocols have been standardized in which an installation should follow. The protocols like siren code for alerting all concerned as set forth by a country are being followed by the industries of the respective countries. OISD in India has approved the following protocols for the oil and gas industries in India as summarized in Table 3.24.

In addition to the above siren codes, the industries should display signage boards in the industry inside roads and plants adequately; i.e., at a regular

TABLE 3.24

Emergency vs. siren code

Emergency	Siren code
Minor fire or gas leakage	½ minute straight run siren sounded.
Major fire	3 minutes wailing siren sounded both in installation and in the colony.
Disaster	3 min. wailing siren-2 min. gap-3 min. wailing siren both in installation and in colony.
All clear	1 min. straight siren after minor, major or disaster.
Siren codes for testing	
Daily test run	1 min. straight run siren at 09-00 hrs.
Wailing siren	2 min. wailing siren on 1st day of month.

distance interval and with good visibility on expected various safe actions from the stakeholders.

Emergencies are classified into the following three categories:

- Emergencies restricted within the industry battery limit and which can be controlled and managed by the industry management with or without support from the state administration.
- Emergencies restricted inside the industry battery limit but cannot be controlled by the industry management themselves and where the control is taken up by the state administration and/or by the respective union ministry of the government with the support of the state administration.
- Emergency spreads outside the industry boundary, going beyond the control of the concerned industry management. The respective state administration takes care of the control with or without support from the union government administration and their facilities.

To support the scenario in the second and third situation above, the union government and state administration should have adequate infrastructures and facilities at their end. They should also form the organogram to take over control of such situation.

3.11 Mock Drill and Management Review of Plant Safety

A mock drill and management review at regular intervals are very important exercises to be conducted to keep operating staff and management in the habit of not forgeting the importance of safety. Actually, these are parts of the safety culture of a petroleum refinery and any other process industry where these are very crucial in addition to following other elements of safety culture like PPE, SOP, and so on.

3.11.1 Mock Drill

In a mock drill, unsafe scenarios like fire is contemplated in a section of the plant; for example, any of the process plants, offsite tank farm, dispatch area, power plant, Effluent Treatment Plant (ETP), flare area, office building, etc. generally, a calendar is made to cover each area over period of time, carrying out the mock drill in one scenario in a month. Overall installation in charge takes the role of review of the performance of the mock drill after the drill is over, while fire and safety department head takes the role of fire chief marshal to conduct the mock drill.

Fire Chief Marshal prepares the calendar and get the approval of overall plant in charge. He also submits the procedure of conducting mock drill to overall plant in charge well before for his approval. The main pre-condition of the mock drill is to ensure that except the overall plant in charge and his hierarchy, nobody in the plant should know that it is mock drill or real scenario.

Case study 1:

One major fire was declared in one crude oil tank in the petroleum refinery. There is already pre-set siren code as mentioned earlier when major fire takes place; there is also pre-set siren code as mentioned earlier when the fire propagates to a disaster. When the fire becomes under control and put off, there is another pre-set siren code as mentioned above to declare the same. The fire chief marshal takes lead in this whole exercise starting from ensuring siren is applied to communication to all concerned through walkie-talkie up to ensuring that fire and safety crew reach the area and fight the fire with prompt response and well prepared while monitoring the alertness of the operation area in charge and his crew and observing the whole activities of all stake holders.

As the fire is a dummy one, operator action may not be real ones but their reflect can be judged by their conversation and body languages, but fire and safety crew should act in real; namely they should turn up to the spot with fire tender immediately, they should start plying fire hoses from various directions and ensure that fire water spray reach up to the top of the tank by ensuring desired discharge pressure of fire water pump. In one case, it was found that water jet was not reaching up to tank top and full tank walls are not wetted by water spray; it was found that discharge pressure of fire water pump came down which the water pump operator could not follow; however, through walkie-talkie communication from field, the same could be established to conclude satisfactory wetting of the tank walls.

Mock drill is then called off with the pre-set yelling of siren followed by review meeting by plant in charge who is called in by fire chief marshal.

In mock drill, not only above achievement, but also participation of every concerned is observed along with their response time and effectiveness of participation.

During mock drill the gate security restrict entry of any employee or outsider without communication from fire chief marshal.

Case study 2:

One major fire was declared in one process plant in petroleum refinery. Following procedure, fire and safety department ensured yelling of siren as mentioned above. The dummy fire was not in a small isolated area such that without stopping operation it could be extinguished rather it called for plant shut down; but as the fire was a dummy one, it was well planned that operating crew will follow mimic actions like using operator console, passing instruction from control room through walkie-talkie to field operators, field operators be in right areas and doing needful actions for proper shutdown of the plant; all the mimic gestures were observed by a group experienced officers as pre-decided.

Mock drill is then called off with the pre-set yelling of siren followed by review meeting by plant in charge who is called in by fire chief marshal.

3.11.2 Management Safety Review

Management safety review is most important element in safety culture of an industry. The review is taken by apex body of the management; namely executive head of the organization. Generally, this review meeting is conducted once in three months. The participation in the meeting consists of the following:

- All functional executive heads or directors.
- All departmental heads from operation including fire and safety, maintenance and inspection, technical services, commercial services, internal audit, administration, medical head, estate office head (if any), site head on occupational health and corporate communication.

The meeting calendar and agenda is prepared by head of fire and safety department subject to approval by operation head to whom he reporting.

In line with agenda, all reports including status reports like minute of minute (MOM) of last meeting which should include all status reports on statutory obligations, mock drills, all highlights on near-misses reports, all incidents reports, and all other pending actionable points.

The meeting agenda also should include all details on occupational health and environmental control issues as detailed below:

- Occupation heath compliance report on employees and pending actionable points.
- Report on statutory obligation on occupational health and pending actionable points.
- Environment control status report and compliance status on statutory obligations.

Hence, from the above, it is evident that the meeting is subdivided into three sub meeting; namely

- Safety
- Occupational head
- Environment

Thus to systematise smooth conduct of all sub meetings at the same time to ensure all heads to keep attention to day to day business, one or two sub meetings can be held in successive days, and also, the non relevant person (s) may be waived out from attending in one or two sub-meetings.

While departmental head from fire and safety spearheads in conducting the safety meeting, site occupational in charge spearheads the meeting on occupational health, and technical service head in responsibilities on environment control spearheads the meeting on environment control review to facilitate apex head of the organization conduct all the sub meetings successfully.

To be mentioned specially, status on training on safety and environment control are given utmost importance.

The meeting must start with safety talk, and after conclusion of the meetings MOM should be issued to all concerned at the earliest.

3.12 Disaster Management Plan (DMP)

The government of India enacted the Disaster Management Act in December 2005. The act provides for the effective management of disasters and for matters connected therewith or incident thereto. The act has a total of 79 sections in eleven chapters.

Various definitions such as disaster, disaster plan, authorities, committees and so forth are provided in the act. The definition of disaster is – a catastrophe, mishap, calamity or grave occurrence in any area, arising from natural or manmade causes, or by accident or negligence which results in substantial loss of life or human suffering or damage to, and destruction of, property or damage to, or degradation of, environment, and is of such a nature or magnitude as to be beyond the coping capacity of the community in the affected area.

Disaster Management means a continuous and integrated process of planning, organizing, coordinating and implementing measures which are necessary or expedient for:

- Prevention of danger or threat to any disaster
- Mitigation on reduction of risk of any disaster or its severity or consequences
- Capacity building
- Preparedness to deal with any disaster
- Assessing the severity or magnitude of effects of any disaster
- Evacuation, rescue and relief
- Rehabilitation and reconstruction

The disaster management plan (DMP) can be summarized to include the following elements:

i. DMP organogram where all area in-charges, including the apex management authorities, should appear with roles and responsibilities.
ii. DMP should include in details the local fire brigade, district authorities, local police station, and local civic bodies as external coordinators with roles and responsibilities.
iii. DMP should include two or more assembly points for emergency evacuation of various offices and sites during disaster, whether a real or mock drill, as approved by a competent authority.
iv. DMP should include conducting two mock drills which should be conducted on an annual basis. The mock drills are mentioned as below:
    - Onsite disaster mock drill which should be conducted for disaster inside the installation arising out of an accident that occurred inside the installation under consideration.
    - Offsite disaster mock drill which should be conducted for a disaster outside the installation arising out of the accident inside the installation.
v. There may be a scenario where an accident inside the installation may result in disaster both inside and outside the installation; i.e., both an onsite and offsite disaster taking place simultaneously or sequentially out of an accident inside the installation. Hence, it is wise to conduct both the drills together if manpower and priorities can afford to do the same.
vi. The mock drills should be conducted as per OISD standards.
vii. The mock drill should be concluded through a meeting involving all concerned and, conventionally, the media should be invited to participate in that meeting.
viii. The apex authorities in the government should be informed prior to conducting the mock drill.
ix. The report should be sent to all concerned.
x. The installation should make action plans to implement the recommendations of the report in a timebound manner.

3.13 Loss Control Manual–A Reflection in Organization Safety

All the aspects of safety discussed so far are the objectives to be fulfilled to ensure loss control management of an organization; however, documentation of the subject separately as 'Loss Control Manual" is necessary for ready reference. The following should be elements in a Loss Control Manual:

i. SHE (safety, health, occupational) policy.
ii. General safety rules of the organization.
iii. Loss control objectives.
iv. Brief description of process units.
v. Safety management systems that are followed.
vi. Leadership and administration.
vii. Leadership training.
viii. Planned general inspection.
ix. Critical task analysis.
x. Accident/incident reporting and investigation.
xi. Task observation.
xii. Emergency preparedness.
xiii. Work permit system with format.
xiv. Accident/incident analysis.
xv. Knowledge and skill training.
xvi. Personal protective equipment (PPE).
xvii. Health and hygiene.
xviii. System evaluation.
xix. Procedure for emergency and change management.
xx. Personal communication.
xxi. Group communication.
xxii. General promotion.
xxiii. Hiring and placement.
xxiv. Materials and service management.
xxv. Off-the job safety.
xxvi. Total productive management (TPM).
xxvii. Six sigma.
xxviii. Integrated management system (IMS).
xxix. Implementation of ISO-14001 (EMS).

xxx. Implementation of OHSAS-18001.

Note: the elements from sl. no. viii to xxvii are the elements in the international safety rating system (ISRS) and ranking is provided to the organization by the competent authority based on scoring on each element. As per ISRS protocol, there are a series of questions against each element which are to be answered for compliance to receive the scores and finally the ranking.

3.14 Hygiene and Ergonomics

As far as hygiene and ergonomics for the employees and all stake holders are concerned, the subject becomes part of the occupational health and safety (OHS). Occupation and human health are interdependent; i.e., to determine the health of the employee without knowing his/her occupation ensures the wrong outcome. Occupational health includes:

- Working methods and working environments that may lead to illness.
- Adjusting to work with an injury or when in ill health.

Occupational health is a balance between the employee and his/her work. It is not only concerned with the diagnosis of disease and its treatment but also with promoting occupational health improvement as a result of the full-fledged program undertaken by the organization, as detailed below:

- Providing a first aid medical center within the industry boundary area.
- Establishing a standard—which may be small hospital—with adequate beds for indoor treatment as well as a facility for outpatient door (OPD) treatment within 5 to 10 KM of the industry. If not possible though required, then the organization has to tie up with the public hospital , located in close proximity of the industry to ensure indoor and outdoor treatment to their employees.
- Having a medical doctor with specialization in industrial health.
- Establishing good hygiene and ergonomics practices under the stewardship of specialist as mentioned above.
- Establishing a safety department that, along with the medical specialist, educates the employees and control their work practices with respect to hazards such as exposure to toxic atmosphere,

oxygen deficient atmosphere, ergonomics, falling from heights, use of PPE (personnel protective equipment), safe behavior during work, health monitoring of employees by keeping records, etc.

3.14.1 Hygiene—Knowledge and Practices

The work environment may lead to the development of diseases like hypertension, diabetes, respiratory diseases, intestinal diseases like chronic stomach upsets, ulcers, upsets in loco motor system, psychological disorders and the like. Hence, in addition to treatment of the diseases, environmental factors should be taken into account and the organization should adhere to good standards to achieve the goal of having healthy employees directly or indirectly. Such environmental factors like hygiene and ergonomics are discussed as follows:

3.14.1.1 Exposure to a Toxic Environment

Repeated inhalation of petroleum products which contain aromatics may cause skin irritation and may result in dermatitis, carcinogenic disease, neuron diseases, gastrointestinal diseases, headache, dizziness, nausea and so on. Even some non-toxic vapors on prolonged inhalation may cause simple asphyxiation due to deficiency of oxygen in the atmosphere; substances like carbon dioxide and sulphur dioxide—which are soluble in water—cause upper respiratory tract irritation and substances which are less soluble in water—like chlorine, phosgene, etc.—cause lung tissue irritation. These gases also cause eye irritation.

Particulate matters like coal dust, silica dust, asbestos, cotton dust and the like cause impairment of the respiratory system. Aromatics like benzene which are present in gasoline cause carcinogenic disease. This is why statutory bodies have restricted the allowable limit of benzene in gasoline.

TABLE 3.25

Effect of various carcinogens on human organs

Carcinogen	Affected organ
Benzene, Arsenic	Bone marrow
Radium, Strontium-90	Bone
Benzedrine, Napthyl Amine	Kidney Bladder
Vinyl Chloride, Arsenic	Liver
Arsenic, Oils, excessive exposure to UVR	Skin
Asbestos, Nickel, Arsenic, Radioactive material	Lung
Nickel	Nose

Tetra-Ethyl Lead, which being a carcinogen, was also being added to gasoline till three decades back. From that point forward, its addition and use in the refinery has been banned. Examples of various carcinogens and the affected organ are given in Table 3.25.

The above list cannot be a complete one. As such, for every chemical used in the industry, there are material safety data sheets (MSDS) which the employees must see and decide the acceptability of the same; if there is no alternative, then chalk out precautionary measures to handle the chemicals and atmosphere while working.

Examples of some toxic gases affecting the central nervous system:

Hydrogen sulphide: very risky and may cause instant fatality in exposure beyond the allowable limit. Hence, employees should carry portable H2S analyser while entering any plant area which may cause release of the gas; for example, in any hydro-treatment unit by malfunction of the process and equipment.

Carbon disulfide: it causes narcotic poison and may be released accidentally as a bottom effluent in a Merox unit meant for mercaptan removal from hydrocarbon.

Noise exposure limit:

It is a general experience that working at a high noise level causes not only auditory effects in human beings, but also may result in cardiovascular attacks, gastro-intestinal disorders and psychological disorders. In view of this, a noise exposure limit has been set in the industry at 90 dB (decibel) with exposure for 8 hours; higher levels like 95 dB are allowed for exposure up to 4 hours, 110 dB up to 1 hour and momentary peak allowed up to 140 dB. Hence, the vendors while supplying the equipment to the industry take care of this issue during design.

The operating industry should have noise monitors like gas monitors, as discussed earlier.

3.14.1.2 Use of Quality Drinking Water

Good quality drinking water represents one of the most important hygiene issues in the industry because poor quality drinking water may result in sickness of the bulk of employees at a time causing a threat to sustaining operation. Consumption of poor quality drinking water causes not only a one-time hygiene problem, but also may lead to chronic diseases among the employees, since drinking water is frequently consumed by the employees on a daily basis; as such, people need to consume about 10 to 12 glass of water per day to remove toxins from the body adequately. Hence, the industry has to maintain a system of supplying good quality drinking water on a continuous basis. Also, as a regulatory measure all countries have their own regulations to maintain the quality of drinking water, not only in the industries but also in every sphere of society and have set up a standard on quality accordingly. The World Health Organization (WHO) also campaigns internationally on the use of quality drinking water.

A drinking water quality standard was prepared with the following objectives:

- To assess the quality of water at resources.
- To check the effectiveness of water treatment and supply by the concerned authorities.

The standards are divided into two categories; namely essential and desirable. Essential characteristics include color, odor, taste, turbidity, pH, alkalinity, total hardness, iron, chlorides and residual free chlorines. Desirable and undesirable characteristics include dissolved solids, calcium, magnesium, copper, manganese, sulphate, nitrate, mercury, calcium, fluorides, phenol compounds, arsenic, cyanides, lead, zinc, selenium, aluminium, boron, anionic detergent, chromium, poly nuclear aromatic, mineral oil, pesticides and radioactive material. The standard also provides a list of micro-organisms, i.e., algae and other bacteria and viruses.

3.14.1.3 Healthy Diet and Kitchen Safety

A healthy diet is one of the vital parameters to sustain industry operation while lifestyle is also to be given importance as both go together. Lifestyle is not a matter of organizational control; rather it should be followed according to individual conscience and the industry should promote programs on lifestyle to educate the employees.

To impart focus on a healthy diet, the importance of nutrients in food need to be discussed as detailed below:

- Macronutrients: these include proteins, carbohydrates and fats which are not only required for energy but also for nutrition resulting in the maintenance of good health. The importance of all these nutrients is discussed below to ensure the quality of food/diet supplied by the industries to their employees.
  a. *Carbohydrates:* About 60% of our diet should be carbohydrates, primarily complex carbohydrates like cereals, rice, roti, pulse, vegetables and not simple carbohydrates like sugar products and refined flours noodles and bakery items that don't contain adequate nutrients. Processed foods available in the market, like cornflakes, contain a good quantity of nutrients.
  b. *Proteins:* Proteins are one of the most important nutrients which help in body building in combination with carbohydrates. There are

about nine essential amino acids in proteins which the body cannot synthesize on their own; these are to be supplied through our diet containing adequate proteins. There are two kind of proteins: namely animal proteins and vegetable proteins. Proteins in vegetables are difficult to be assimilated in the metabolism process as compared to animal proteins. On the other hand, there are proteins in animal foods fit for assimilation but there may be lots of fat in the animal food which is not easily digestible. So, our diet should be a balance of vegetable food and animal food. Among the vegetables, pulse and soya bean have very good quantities of protein in addition to carbohydrates. Rice also contains proteins but less in quantity compared to pulse and soya bean but have more carbohydrates compared to them. Vegetables contain mostly carbohydrates but have specific mineral contents which are different from one kind of vegetable to the other; hence, there should always be a mix of various kinds of vegetables in our diet; i.e., if not daily, there should be weekly variation in vegetable variety intake. From infants to the eldery, a balanced food is milk which contains balance in the quantities of carbohydrates, proteins, fats and minerals; hence, if possible, we should take one glass of milk every day, but with age many people face gastro problems that require them to avoid taking milk.

c. *Fat:* We need taste while taking our food, and food containing fats and sugar contribute most in taste; however, as mentioned earlier, fat should be taken in less quantity due to difficulty in digestion. Fat is also an important source of energy; i.e., it is easily combustible as it is basically triglycerides of fatty acids. This means on hydrolysis it generates fatty acid and glycerol, both of which are easily combustible. A diet should include mixed fat, i.e., one teaspoon of ghee, two to three teaspoons of saturated vegetable oils (if one can digest them), along with unsaturated vegetable oils and tinges of creamy food to provide taste.

- Micronutrients: These are required in small quantities. Vitamins and minerals are available in all kinds of vegetables, cereals and animal food but in different proportions and particular vitamins and minerals are available in specific foods of the above. Hence, as mentioned earlier, there should be variation in our diet daily or weekly to ensure we are getting all types of minerals and vitamins for our health and immunity. Fruits also contain good quantities of vitamins and minerals.

Kitchen safety

Wherever there is a kitchen for a mass supply of food, there is a tendency to overlook hygiene parameters in maintaining the food preparation process in the kitchen and upkeeping the ambience of the kitchen. In the industry, generally, a committee/task force is formed from the employee and management sides to oversee the cleanliness of kitchen and its staffs, ambience like wire net protection in the windows and door closures to avoid entry of insects and cattle, along with good housekeeping of the food stores. Some tips to maintain the practice of keeping a good kitchen are given below:

- The storeroom should be refrigerated and pesticides sprayed from time to time to avoid ingress/formation of insects/bacteria.
- If one likes to preserve cooked food, it is to be refrigerated within two hours of cooking, but in the industry, preservation of cooked food is not encouraged.
- Sanitize your kitchen at regular intervals, but never spray any volatile pesticides in the kitchen as this may cause an explosion when putting the burner on.
- Maintain good cleaning of all utensils and preferably use a sterilization process like keeping these on steam or washing these in detergent solutions along with one teaspoon of chlorine bleach.
- Wash the vegetables (before cutting) to clean and remove all pesticides if present; washing after cutting may remove some nutrients.
- Wash mutton and chicken with hot water before taking these up for cooking to kill any bacteria present.
- Cooking temperature should be at least 75 °C to kill germs.
- Cooking in low flames always make the food tasty and it is also healthy—as in this process charring of oils and food items are avoided.
- Don't consume half-cooked items; for example, half boiled eggs which may be harmful due to the bacteria not being killed properly.
- Kitchen staff should use an apron while in the kitchen and they should always keep their hands washed after every phase of cooking. They should also not use shoes in the kitchen; if necessary, they can use sleepers meant only for use inside the kitchen. As such ,nobody should be allowed to enter the kitchen without removing their shoes.

Maintaining good practices in food selection, purchasing and cooking does not end with maintaining good hygiene habits: the industry should also establish a good ambience in serving food as well as a good delivery system where transport of food from kitchen to workplace is necessary.

3.14.2 Ergonomics—Knowledge and Practices

The word, 'Ergonomics' comes from the Greek words, 'ergon' which means work and 'nomos' which means law. Hence, the basic definition of 'ergonomics' should be the various rules to be followed while executing any work. As a subject of study, research and application in industries and daily spheres of activities, it has immense importance and is developed as a subject of science and engineering that focuses on human capabilities and limitations in carrying out work tasks for shorter as well as for longer duration. It deals in the design of tools, equipment and workstations which should take care of human capabilities to execute the work with comfort and without strains and diseases. Ergonomics is based on facts, research, experiments and theories using human anatomy and mechanisms as a base line. Overall, it deals in interactions among human beings, their work and the environment of work.

Ergonomics should be subdivided into the following categories:

- Ergonomics in the control room
- Ergonomics in the laboratory
- Ergonomics in construction
- Ergonomics in day-to-day field operation
- Ergonomics in office work

Hence, while designing tools, equipment, furniture and setting postures in work execution, ergonomics should be considered separately as mentioned above. In fact, guidelines of the 'International Ergonomics Association Executive Council' should be referred to while setting out the implementation procedures.

Principles of ergonomics

Following are the principles of ergonomics which should be pursued in practices, not only in the industries, but also in every spheres of life:

- Work in neutral posture.
- Work at proper height.
- Provide clearance between worker and work object/station.
- Reduce excessive force on body joints, pressure points and muscles.
- Keep everything in reach; i.e., body bending to collect, pulling/pushing nearby objects to be avoided and to reduce excessive motion.
- Minimize fatigue and static load.
- Move out for exercise and stretch for a while after sitting/standing at a fixed posture for a long period; say, an hour.

- Maintain a comfortable environment like proper lighting, air conditioning inside the work room or an exhaust fan during a power failure. Also, provide good air circulation in the field workplace if it is too hot and work for longer duration is required. If it is the rainy season, a proper tent is to be made for working, etc.

Benefits of following ergonomics:

- Safer job without injuries.
- Higher productivity and efficiency.
- Improved qualities of job.
- Physiological and psychological satisfaction.

Ergonomic injuries/Musculo-skeletal disorders (MSDs):

- Can affect muscles, nerves, ligaments, joints, cartilage, spinal cord, etc.
- MSDs is the result of repetitive motions of the above organs.

Ergonomics in sitting chair design and postures:
Postures:

- Shoulder should be relaxed.
- Forearms and hands should be in a straight line.
- Eye should be at a distance of about 45–70 cms from the display monitor.
- Lower back should be supported on the chair backside.
- The top of the monitor screen should be at or slightly below eye level.
- There should be clearance under the PC/DCS monitor table top so that a worker's legs can be put forward inside the table top.
- if a PC or laptop screen is more vertical and less inclined backward, then the human body backside can be slightly leaned to the chair backside; for DCS operation in the control room of any industry, it is observed that the DCS monitor is inclined backward as per industry norms; in that case, it is found that the operator's backside if kept in a relaxed erect position gives more comfort in work.

Design of chairs:
There are already predesigned chairs in the market are available for use for the above purpose. However, the following should be kept in mind while purchasing the design:

- Chair should be cushioned both at the bottom and particularly at the backside because backbones and discs contribute only about 30% weight while tissues and muscles in there contribute about 70% weight. Hence, as tissues and muscles take the most load, these are very soft compared to bones and the chair backside should be cushioned to avoid pressure on the tissues and muscles.
- Chair height should be adjustable so that operator can adjust height as per his/her own height.
- Flexibility of chair backward should be possible but be restricted up to a small degree to avoid the operator making a mistake and having an accident.
- Chair quality, strength and weight should be as per standards in the respective countries.
- For desktop PC or DCS monitor, provision in the table design should be provided to keyboard on a separate movable rack just below the table top where the monitor is installed, instead of keeping the keyboard in the table top.

Self-evaluation and checklisting of ergonomics issues:

Working people are the most important resources to utilize feedback to provide the right ergonomics and improve it further in the workplace. An attempt has been made to develop these two in formats. Industries should also incorporate their own experiences to improvise the formats because it should be mentioned that ergonomics is ultimately to facilitate good health of the workers in the long run and to obtain satisfaction/comfort to work for longer periods without being distracted.

Self-evaluation questionnaire:

- Do you know how to adjust the chair?
- Does your chair have height, back and rotating adjustment facilities?
- Are your legs firmly supported on the floor?
- Does your chair have adequate cushioning?
- Is your chair of the right strength to feel secured while sitting?
- Is sitting comfortable and quality up to satisfaction?
- Are your shoulders relaxed?
- Are your elbows next to your sides?
- Are your forearms in straight lines to the keyboard:
- Are your forearms properly resting on chair side handles?

- Is the mouse at the same height as the keyboard and placed next to it?
- Do your wrists bend while using a keyboard or mouse?
- Is your head straight and not bent while viewing the monitor or using the keyboard?
- Are frequently used items in close proximity?
- Do you follow macros, short keys to avoid repetitive typing?
- Do you take bio-breaks from time to time during your work?
- Have you understood the work ergonomic principles and practices to be followed?

Checklist questionnaire:

- Is lighting adequate and without glare?
- Is the PC/DCS monitor at the right position with respect to lighting?
- Is proper ventilation or proper air conditioning maintained?
- Is layout of the area appropriate?
- Is access to work area free from obstacles?
- Is access area free of trip or slip hazards?
- Is there adequate space, i.e., drawers below the table to keep regular usable items?
- Is there daily usable items listed beforehand to keep within reach?
- Is area free from noise?
- Is there a tool available to access shelves if shelves are required?

3.15 Safety Audit

Whether the industry is following all the safety prerequisites as mentioned above on a sustainable basis should also be reviewed through auditing over a period of time. The more the hazards in the industries, the more the importance of doing a safety audit. It is compulsory in the petroleum industries as per OISD (Oil Industries Safety Directorate) guidelines.

Safety audits are of two categories; namely Internal Safety Audit (ISA) & External safety Audit (ESA). In ISA, audit team members comprise in-house employees of the organization if the organization has adequate manpower with expertise for doing the audit. Otherwise, they hire external authorized agencies for the same. In ESA, audit team members by compulsion are external to the organization where the audit is to be carried out. In Indian petroleum industries, OISD nominates audit team members having adequate operational experience in the same industries. It is possible this may be someone the team member worked with in the same organization earlier but presently his status has changed to become an official of OISD. For other industries in India, external team members are selected by the offices under jurisdiction of respective ministries of the government. ESA is customarily conducted every alternate year. Hence, ISA is conducted in the year in-between.

3.15.1 ISA

Methodology adopted to carry out the audit:

i. Visit to process units and utilities.
ii. Study of operation manuals, P&ID, and layouts of the units.
iii. Interview of personnel involved in operation and maintenance of the process units, off sites and utilities, store keepers in materials department and technicians in workshop and garage.
iv. Study of the fire and safety manuals and safety management system manual.
v. Checking the incident/accident reports.
vi. Physical condition tour of the industry from process units to offsite, including effluent treatment plant, firefighting facilities, storage and dispatch, power and utilities units etc.

Note: for petroleum refineries, there is a ready reckoned checklist available in OISD-GDN-145 which should be followed in carrying out ISA.

Content of the report:
The report should contain the following:

i. Observations and recommendations.
ii. Detailed findings as per stipulated guidelines, e.g., for petroleum refineries, as per OISD 145 guidelines.
iii. Physical condition checks.

Observations and recommendations as in sl. no. (i) should contain the following:

Noteworthy observations.

Observations which call for improvement.

In detailed findings as in sl. no. ii), all detailed observations should be listed out for each process unitwise for each subsection of offsite; namely tank farm, effluent treatment plant, etc., each dispatch section; namely tank truck dispatch, tank wagon dispatch, pipeline dispatch, if any, pressure vessel dispatch, etc.

The observations should be documented in format and in simple English for the purpose of good communication. There should be three columns in the format; namely sl. no., description of item/area audited, and auditors' observations.

Area wise detailed finding in audit of petroleum refinery as a case study:

I. Process units:

i. Layout and general:
Audit of general areas are done like layout and drainage as per OISD-118 where approach roads to plant, oil water sewers are adequately sized and adequately covered or not, ladders/platforms are provided as per standard engineering practices or not, surface water drainage is adequately sized or not, whether closed blow down (CBD) system for draining of hydrocarbon from the plant equipment have been provided or not, etc.

ii. Fire protection facilities:
Audit is done as per OISD-116 and the following are checked:

- Design of fire water network is checked to see whether its stipulated design of ring type and covering all sides of each facility and plant is maintained and accessible or not,
- whether the hydrants/monitors are provided from all directions and at a safe distance of 15 meters from nearest hazard or not,
- whether steam lancers are provided at all possible leak sources (as per visual inspection) or not,
- whether adequate numbers of portable fire extinguishers are provided in all plants or not,
- whether communication systems, fire alarm and public announcement system are working or not,
- whether copies of firefighting manual are available or not,
- whether safety committee meetings at plant level are conducted or not.

iii. Housekeeping:

- Whether floors of all plants and equipment areas are properly washed or not to avoid oil accumulation,'
- Whether all plants areas are in cleaned condition to remove all debris,

- To conclude general housekeeping is adequate or not.

iv. Training:

- Whether copies of updated operating manuals, fire and safety manuals, disaster control manuals, standing instructions are available or not,
- Whether training needs for the employees have been identified or not and training have been provided or not accordingly,
- Whether contractor employees training programs are made or not and the same are being followed or not,
- Whether employees and workers are using proper personal protective equipment (PPE) or not.

v. Inspection and maintenance:
Audit is done as per OISD-129 & 127 and following are checked:

- Whether inspection of equipments/tanks/piping/safety valves/steam boilers, etc., are carried out as per schedule or not,
- Whether all modifications are done as per approved engineering drawing or approved management of change (MOC) or not,
- Whether well designed work permit systems as per OISD-105 are being followed or not, for e.g., permit for hot/cold work, permit for energizing and de-energizing, permit for excavation, permit for vessel entry, permit vehicle entry inside the plant, etc.

vi. Electrical:

- Whether all motors, fitting are as per electrical area classification or not and whether area classification drawing or not and area classification drawings are available or not,
- Whether illuminations are adequate in all facilities and in the plant,
- Whether equipments earthlings are adequate or not,
- Whether emergency lighting system is provided or not,
- Whether any interlock is bypassed or not, and the same is recorded, approved or not.

vii. Instrumentation:
- Whether electronic instrumentation is provided or not as per area classification,
- Whether minimum safety instrumentation in furnace, vessels, columns, heavy duty pumps and compressors are provided or not,
- Whether any schedule exists for routine checks of the interlock systems or not,
- Whether all instrument analyzers are checked regularly as per schedule or not,
- Whether interlocks are bypassed or not, and the same are recorded, approved or not,
- Whether smoke/gas/heat/hydrocarbon detectors are provided and working or not.

viii. Process subsystems:
- Control room:
  a. Whether fire protection is provided and adequate or not,
  b. Whether P&IDs are available or not,
  c. Whether emergency shutdown control systems are available or not,
  d. Whether PPE and first aid kits are available or not,
  e. Whether control is blastproof or not,
  f. Whether control room air condition system is properly working or not,
  g. Whether cable entry in control room is properly sealed or not,
  h. Whether inert gen system inside the control during fire, if any, is provided or not
- Pump house:
  Audit is done as per OISD-118 and check points are as below:
  a. Whether pumps are located at a safe distance from furnace or not,
  b. Whether hot pumps are segregated from pumps using light hydrocarbon or not,
  c. Whether pump mechanical seal systems are as per API requirement or not,

d. Whether pumps are easily accessible and piping are not fouling the access or not,
e. Whether hot pumps are properly insulated or not,
f. Whether hot pumps are provided with warm up lines and the same are in use or not,
g. Whether vents/drains of hot pumps are connected to cooling system or not,
h. Whether reciprocating pumps are having proper safe release system or not,
i. Whether all threaded connections are healthy and not corroded or not,
j. Whether drains are connected to CBD or OWS or not as needed in respective cases,
k. Whether all OWS funnels are through and not choked or not,
l. Whether auto start/stop of pumps where applicable checked as per schedule or not,
m. Whether pumps running hours are recorded or not,
n. Whether lube oil cups in pump bearing house are properly filled in and always being maintained healthy or not.

- Compressor house:
  Audit is done as per OISD-118 and check points are as below:
  a. Whether located at safe distance from furnace or not,
  b. Whether location is well ventilated or not,
  c. Whether the following are in working conditions or not:
     - vibration monitoring system
     - anti surge device
     - speed control system
     - shutdown system
     - lubrication system

  d. Whether alarms and signaling systems are working or not,
  e. Whether safety valves are checked or not,
  f. Whether suction strainers are periodically cleaned or not,
  g. Whether compressor has proper de-pressurizing system or not,

h. Whether condensable collection system at suction/discharge is working or not,
i. Whether compressor has nitrogen inerting system or not,
j. Whether stuffing box vent is through open to atmosphere or not; otherwise as alternative inert gas sealing along with vent to flare are provided or not.

- Heat exchangers/condensers/coolers:
  a. Are adequate numbers of high point vents (HPV)/ low points drain (LPD) with proper isolation valves is provided?
  b. Whether provision of thermal relief/expansion is in place or not,
  c. Whether insulation/cladding is fully covering the exchangers or not,
  d. Is earthling provided to each exchanger?
  e. Are supports fireproofed?
  f. Do sacrificial anodes exist for coolers/condensers where different metals are in contact?
  g. Whether the operating parameters and test pressures are tagged on the equipment and are being operated within?
  h. Whether the overhead condensers platform is within the reach of firefighting and having two sets of staircases or not.

- Process vessels and columns:
  Audit is done as per OISD-152:
  a. Whether the relief valves are periodically tested and record maintained?
  b. Whether all utility connections are in working condition?
  c. Whether the vessels/columns have been provided with adequate safety instrumentation?
  d. Whether all nozzles, nipples, vents and idle drains are capped and blinded?
  e. Whether the vessels/columns supports are fireproofed?
  f. Whether insulation/painting are proper?
  g. Do all vessels/columns have two sets of level indicators?

- Process furnaces:
  Audit is done as per OISD-111:
  a. Whether pass flows are provided with controller and recorder and are working?

b. Whether skin and bride temperature are recorded?
c. Whether main burners are provided with pilot burners?
d. Whether pass flows low alarm for warning and low-low alarm for cutoff fire have been provided?
e. Whether high skin and bridge temperature alarms have been provided?
f. Whether furnace cutoff /trip system is provided, covering all other safety aspects like low combustion air pressure, high box pressure in case of closing of stack damper, if any and low-low feed flow?
g. Whether periodical checking of trip system is done?
h. Whether furnace has step ladder and monkey ladders to reach each platform?
i. Whether safety wears such safety goggles, face shield are used by the operators for viewing furnace inside?
j. Whether explosion doors are properly placed?
k. Is snuffing/purging steam manifold easily accessible?
l. Is the coil purging steam facility is in line with OISD-111 annexure-1?
m. Whether stack damper control system is working satisfactory and stack damper opening indicator is readable from ground level?
n. Whether control system of drop out door opening can be tested on line?
o. Whether location and height of FD fan air intake can be considered safe?
p. Whether the furnace has decoking system if needed and operating?
q. Whether noise level near the furnace is within limits and monitored at least once a year?
r. Whether condensate draining, OWS connection and manhole location can be considered safe?
s. Whether support structures are properly fireproofed?

- Process piping:
  a. Whether expansion loops/expansion joints are working?
  b. Whether TSVs provided are periodically tested, records maintained?
  c. Whether pipe supports (sliding/spring supports) guides and anchors are in order and unlocked?

d. Whether piping slopes are towards drain points?
e. Whether piping above 60^0 C above surface temperature are having safety insulation?
f. Whether continuity jumpers are provided between flange joints for LPG, Naphtha, ATF, etc. (as per OISD-106)?

- Relief and disposal system:
  a. Whether relief valves have set at 10% above normal operating pressure?
  b. Whether difference between set pressure and normal operating pressure is maintained minimum 2 kg/cm^2g?
  c. Whether safety valves installed directly on the protecting equipment/on the connected piping without any isolation valve if it is single safety valve?
  d. Whether multiple safety valves have been used? If so, then whether set pressures are same or if different whether there is no isolation valve and directly connected to protecting equipment?
  e. If multiple safety valves at same operating pressure to keep one as standby and isolation valves are used in that case, whether one valve is at least in lock open condition and valves are vertically downward mounted to avoid valve gate of standby safety stuck up in closed condition?
  f. Whether any relief valve on hydrocarbon/toxic gas services relieves to atmosphere? If so, whether the following points are considered?
     - Discharge vertically upward through individual stack.
     - Minimum exit velocity of 15 m/second.
     - Snuffing steam provided.

  g. Whether use of common atmosphere vent stack for several relief valves is observed?
  h. Whether any vent stack is observed discharging at an elevation of less than 3 meter above the tallest structure within 15 meter radius?
  i. Whether acoustic insulation has been provided on vents producing excessive noise?
  j. Whether toxic vapors like SO_2, Phenol, Chlorine, etc., are disposed through closed system?
  k. Whether isolation valves with spectacle blinds have been provided at the batter limits of the plants? If so,

whether their valve stems are pointing downward to avoid accidental closure?

l. Whether unit knock out drum has been provided? If so, can it hold discharge for 5–10 minutes from a single contingency?

m. Whether pumps with auto start/stop at high/low levels have been provided to pump out the content of knock out drum?

n. Whether high/low level alarms have been provided in the control room?

o. Whether vapor depressurization system has been provided for emergency situations such as runaway reaction, exposure to fire, etc.?

- First aid:

  a. What is the percentage of shift employees trained in first aid? How many provide artificial resuscitation?

  b. Does every control including pump houses possess a first aid box, water gel blanket and explosive meter?

  c. What is the accident record of the last two years?

II. Crude oil and product tank farm:

i. Layout and general:

a. Whether the location, layout, interspacing in the tank farm meet OISD-118/statutory requirement?

b. Whether dyke/fire walls meet the above requirement?

c. Whether the roads are available around the tank farm for movement of fire tenders?

d. Whether piping in the dyke farm is kept to a minimum?

e. Whether piping not connected to a tank is passing through the dyke?

f. Whether any hydrocarbon detection system is provided? If so, whether authorized evaluation was made before installation and working satisfactory?

g. Whether housekeeping is satisfactory?

h. Whether communication system is available and working?

i. Whether illumination in the area is adequate?

j. What is the procedure adopted for recovery and disposal of tank bottom sludge?

ii. Sewers:

a. Whether the tank farm closed and open drains are clear?

b. Whether provision exists to divert the tank farm drainage either to storm water or to oily sewer?

iii. Tanks:

a. Whether tanks are properly painted and important specifications are marked?

b. Whether nozzles of 2″ and above have been provided with reinforcement pad and with tell-tale holes?

c. Whether the tanks are provided with circumferential railings?

d. Whether roof drains of floating roof tanks are provided with check valves?

e. Whether alternative arrangement to read/record levels has been provided?

f. Whether double level indications have been provided?

g. Whether any product is stored above its flash point in fixed roof tank?

h. Whether any product is stored above its flash point in floating roof tank? If yes, what precaution has been taken?

i. Whether the breather valve provided matches the pumping in/out requirement?

j. Whether any settlement is observed in the tank pad?

k. Whether tanks on ATF and chemicals are fully coated internally?

l. Whether steam coils provided have any flange connection inside the tank? Whether provision exists for leak detection in steam coils?

m. Whether fixed/semi fixed foam system is provided on all tanks storing A class and B class products (18 M and above for floating roof) and also tanks storing C class products of more than 40 meter diameter?

n. Whether sprinkler system has been provided on the tanks as per OISD norms?

o. Whether Halon system has been provided on any tank? If so, whether periodically Halon system is checked and record maintained?

p. Whether tanks are properly painted and numbered?

q. Whether flame arrestor is provided in fixed roof tanks which are likely to have explosive mixture in the vapor space?

r. Whether goose neck type vent connections have been provided with suitable wire mesh to avoid birds entry/formation of nest?

s. Whether oil of proper viscosity is used in oil breather valves?

iv. Piping:

a. Whether pipe alleys are properly graded to ensure that no oil/water gets accumulated below?

b. Whether pipe supports, anchors, guides are in order?

c. Whether pipes carrying heavy properly heat traced and insulated?

d. Whether proper color coding, product identification marks have been provided?

e. Whether low point drains and high point vents have been provided with proper caps?

f. Whether hammer blinds are provided in manifolds and are operable?

g. Whether floor under the manifold is paved, have curb walls and directed to OWS?

h. Whether steam traps of tracer lines are working?

i. Whether any short bolt is found or bolt is missing on flange?

j. Whether any redundant line is kept positively isolated with all ends plugged/blinded and actions are taken to get them removed?

v. Operating procedure:

a. Whether a copy of an updated operating manual with all operating instructions, blending procedures is available with all operating personal and a copy is available in the control room?

b. Whether any chance of product contamination during tank line or product changes over? If so, whether tank change over schedule has been properly defined?

c. Whether periodical checks are made for locating leakage, if any, on tank, floating roof valves, piping, etc., and recorded for rectification?

d. Whether roof drain valves of floating roof tank are kept open? If no due to roof leakage, whether during rain valves are immediately opened and recorded and tank is marked for outage at the earliest?

e. Whether cross-checking of auto level gauge is done through manual gauge while the tank is in operation?
f. Whether it is ensured during tank filling in that a maximum velocity of 1 meter/second is maintained for products like SKO/ATF?
g. Whether the use of nylon rope for samplings was observed?
h. How it is ensured that side entry mixers are working when liquid levels are above the blades?

vi. Inspection and maintenance:

a. What are the schedules of inspection of tanks as per OISD-129 and whether the same is being followed and recorded?
b. What is the frequency of PV vents, relief valves, breather valves, flame arrestors? Whether records are maintained?
c. Whether any provision exists for static electricity discharge? What is the frequency of earthing connections inspection and whether the same is recorded?
d. Whether pontoon covers are in place and are in good condition?
e. Whether foam/neoprene seals are checked regularly for perfect sealing and record maintained?
f. Whether any documented procedure exists for tank cleaning? What is the methodology to ensure strict adherence?
g. Whether the ladders and staircases are in good condition?
h. Whether the pipeline relief valves are relieving to tank or to outside?
i. Whether inspection is carried out to repaired tank before it is taken into operation?
j. Whether inspection schedule of radar gauges provided on storage tanks have been followed?
k. What the schedule of painting of tanks and whether the same is being followed?
l. Whether nitrogen blanketing system is provided in any tank? If so, what is the frequency of inspection and whether record is being maintained?

III. POL Loading and unloading—tank trucks:

i. Headers and loading hoses:

a. Whether proper arrangement has been provided for securing the hose nozzles with tank manhole?

b. Whether facility is provided to drain an overfilled truck to a closed system with proper pump out arrangement?

ii. Placement of truck:

a. Whether any system exists to regulate the entry of trucks, filling and releasing?

b. Whether a truck parking place has been provided both for empty and filled trucks?

c. Whether any safety check is carried out in trucks before issuing loading advice?

iii. Loading operation:

a. Whether it is ensured that that truck loading hoses are properly connected to earthling system before commencing loading operation?

b. Whether decantation of any overfilled truck is done to a closed system and with close supervision?

c. Whether following are ensured after completion of loading?

- fill pipes are properly closed.
- top covers are properly closed and sealed.

iv. Vehicle and crew check:

a. Whether the vehicle design has been approved by CCE?

b. Whether the license and explosive certificates are available in the vehicle?

c. Whether the vehicle is in good condition and RTO has issued a fitness certificate?

d. Whether the vehicle has non-sparking electrical fitting and external wiring through conduit?

e. Whether the vehicle has a cutoff master switch for electrical system?

f. Whether there is facility to shut off the drain valve during an emergency?

g. Whether the vehicle has two fire extinguishers?

h. Whether the gross and tare weight painted on the vehicle tally with the permit?

i. Whether the vehicle has a first aid box, tool box and emergency lighting?

j. Whether there is any source of ignition like candles, match box, etc., in the vehicle?

k. Whether the vehicle has at least one driver and one cleaner as its crew member?
l. Whether the driver has a valid driving license?
m. Whether the crew is trained to handle hazardous nature of petroleum products, emergency procedures, emergency communication, etc.?
n. Whether the driver has a copy of standing instruction and TREM card?
o. Whether the exhaust is wholly in front of the tank truck and has ample clearance from the fuel oil system and combustible material?
p. Whether approved quality spark arrestor provided on the engine exhaust and muffler/silencer is not cut off from the exhaust?
q. Whether each compartment of the tank is fitted with independent PV vent with a minimum opening of 3 sq. cm.?
r. Whether vent openings are covered with two layers of wire mesh of size not less than 11 mesh?
s. Whether the tank truck is conspicuously marked on each side and rear, in bold letters at least 7 cm high and on a background of sharp, contrasting color as per UN code?
t. Whether effective spark arrester is fitted on air intake of engine?
u. Whether electric wiring is of heavily insulated quality and adequately carry maximum load?
v. Whether suitable over current protection in the form of fuses/automatic circuit breakers, etc., have been provided?
w. Whether all junction boxes are sealed properly?
x. Whether electrical equipment like generator switches, fuses and circuit breakers are located inside the cabin or engine compartment?
y. Whether batter is in easily accessible position with a heavy duty switch close by, to cut off the battery in emergency?
z. Whether battery terminals have protective rubber covers?
aa. Whether vehicle cabin is checked for the presence of any flammable/explosive substance being carried by the crew?

IV. LPG dispatch (storage/loading and unloading):

Bulk storage area:

i. General:

i. Whether the layout, area classification drawings have been approved by CCE?
ii. Whether any new facilities have been installed? Are they safe to use?
iii. Whether general housekeeping is satisfactory?
iv. Whether storage vessels are insulated? If so, what is the condition of insulation?
v. Whether five year hydraulic test records are available?
vi. Whether any sign of corrosion in the pipelines? Whether painting has been done as per schedule?
vii. Whether there is any short bolt and all bolts of proper sizes have been used and no bolt is missing?
viii. Whether any portion of pipe length found buried with soil or undergrowth?

ii. Relief system:
a. Whether a relief system is adequately designed?
b. Whether relief valves are open to flare/atmosphere? Any steam connection given if system is open to atmosphere?
c. Whether relief valves are provided with isolation valves? If so, what is the system/method to avoid inadvertent closing of isolation valves?
d. Whether the safety valves are periodically tested?

iii. Draining:
i. Whether water drain chamber has been provided away from shadow of sphere and whether the valve for draining is away from the drain opening, if open drained?
ii. Whether the area under the sphere has been provided with concrete pavement with proper slope?
iii. Whether antifreeze valve provided on drain line?
iv. Whether double block valves provided on water draining line (quick shut off type near the chamber and the other a globe valve)?
v. Whether safety instructions for draining operation have been prominently displayed?
vi. Whether draining is done under constant supervision?
vii. Whether sampling points provided with duplicate valves?

iv. Instrumentation:

a. Whether level gauges of two types have been provided for direct level measurement and counter checking?

b. Whether monitoring of levels of all the vessels is being done? Records are available?

c. Whether high level alarms have been provided in storage vessels and are in working condition?

d. Whether remote operated valves (ROV) have been provided on liquid outlets/inlets and are checked for proper operation?

v. Fire and safety:

a. Whether a gas detection system is provided in the area and is working satisfactory?

b. Whether the capacity of fire water pumps and header are adequately designed?

c. Whether sprinkler/deluge systems are provided on all vessels and are periodically tested for satisfactory operation?

d. Whether sprinkler/deluge valves are remote operated and are located 15M away from storage vessels?

e. Whether supports of the vessels are fireproofed?

f. Whether periodical checks are done with explosive meters to detect any gas leak?

vi. Electrical:

a. Whether bonding strips are used across the flanges for electrical continuity?

b. Whether all electrical fittings in the area meet electrical area classification?

c. Whether earthing connections are provided and are properly maintained?

d. Whether illumination in the area is adequate?

vii. Pumps/Compressor area:

a. Whether layout of the area meets statutory requirements in terms of safety distances from storage vessels, roads and other facilities?

b. Whether the pumps are having double mechanical seals to minimize leakage? Are the seals in good condition?

c. Whether a recirculation line to storage vessel provided?

d. Whether venting is done through a high rise vent? Whether steam connection is provided to atmospheric vent?
e. Whether drains/vents/sample points have double isolation valves and are provided with caps?
f. Whether a gas detection system has been provided in the area and is working satisfactory?
g. Whether all motors and other electrical equipment such as terminal/junction box, lighting fittings, etc., are weatherproof/flameproof and as per hazard area classification requirements?
h. Whether the motors are provided with proper earthling?
i. Whether illumination is satisfactory?
j. Whether all equipment is properly painted?
k. Whether established communication system exists between the loading area and pump area during normal operation/ emergency?
l. Whether safety instructions have been displayed?
m. Whether maintenance tools used are non-sparking?
n. Whether long range monitors/sprinkler system have been installed to cover the area and are in good condition?
o. Whether adequate numbers of fire extinguishers have been provided?

V. Tank wagon gantries:

i. Gantry area:

a. Whether layout of the gantry meets statutory requirements?
b. Whether the design, material selection, installation of gantry platform, loading header, loading hoses, etc., have been done as per standard engineering norms adopting proper safety guidelines?
c. Whether facilities provided to handle leaky wagons in situ (evacuation of tank wagons)?
d. Whether emergency stop push button for loading pumps exists in the gantry?

ii. Loading headers and arms:

a. Whether loading arm swivel joints' gas seals are periodically greased and maintained properly?
b. Whether quick isolation valve/MOV has been provided on the loading header and is operative?

c. Whether independent isolation valve provided on each loading point?
d. Whether loading velocities are maintained within safe limits by providing restriction orifice in each point?
e. Whether loading headers are provided with proper earthling/bonding connections?
f. Whether loading hoses not in use are stored properly and open end kept blinded?
g. Whether the following facilities are available on each loading point?
   - excess flow check valve on the fitting line,
   - vapor return line with check valve back to storage tanks, Flexible hoses on filling vapor return line.

h. Whether a depressurizing system provided in the gantry header is discharging to high point vent/flare?
i. Whether temperature and pressure gauges are provided at convenient points?
j. Whether loading is provided with ISV relieving to high point vent?
k. Whether proper capacity weigh bridge is provided on each spur of the gantry? Any risk of gas accumulation in the pit?
l. Whether bonding strips are provided for continuity of rail tracks?
m. Whether bonding connectors (copper strips/wires) are provided for piping?

iii. Fire protection and safety:
   a. Whether a water sprinkler system is provided and working satisfactory?
   b. Whether adequate numbers of fire hydrants and large range monitors have been provided?
   c. Whether gas detector system is provided with an alarm in control room?

iv. Operating procedure and facilities:
   a. Whether an updated operating manual with loading/unloading procedures is available in the control room and is well understood by operating crew?
   b. Whether the TXR staffs (Railway concerned technical staff)

check and issue fitness certificates in writing for fitness of tank wagons?

c. Whether checks are done for gaskets, 'O' rings w.r.t. physical damage, if any, any corrosion on tanker, safety valve, thermometer, etc., before commencement of loading operation?
d. Whether the following are ensured before loading operation?
    - all engines nearby are switched off,
    - all hot jobs in the area are stopped,
    - Proper earthling connections made,
    - All wagons are secured by hand brake.
e. Whether it is ensured that non-sparking tools are used for connecting/disconnecting hoses?
f. Whether use of any synthetic fiber material was observed?
g. Whether loading hoses are periodically hydro tested, discarded/replaced and records maintained?
h. Whether all hoses carry test code and date?
i. Whether constant supervision is done during loading/unloading operation?
j. For detailed checklist for loading T/W, whether OISD-144 is followed?

VI. Tank truck gantry:

i. Facilities:

a. Whether the truck loading facilities have been installed as per CCE approved drawings?
b. Whether separate entry/exit gates exist for the tank trucks?
c. Whether Weighs Bridge is available within the installation and is calibrated periodically and maintained?
d. Whether the number of trucks parked at the gantry area is restricted?
e. Whether a facility is available for evacuation/decantation of defective/leaky trucks with necessary high point vents and with steam connection?
f. Whether the following random checks are done for tank truck?
g. valid licenses,
    - Fittings as per CCE such as flame arrestor,
    - Fire screen, etc.,
    - Safety instruction booklet.

h. Elimination of ignition sources like loose wires, uncovered battery terminals, etc., are ensured?
i. Whether a gas detection system is provided in the area and working satisfactory?

ii. Operation:
a. Whether the vehicle engine is always shut off during loading operation?
b. Whether it is ensured that the earthling connection is properly checked before connecting hoses to tank truck?
c. Whether blank flanges are fitted on the truck inlet flange and to the end of hoses not in use?
d. Whether loading arm is open and blinded when not in use?

iii. Fire protection safety:
a. Whether water sprinkler system is provided and working satisfactory?
b. Whether adequate number of fire hydrant and long range monitor have been provided?
c. Whether a gas detection system provided with alarm in control room?
d. Whether provision to start water sprinklers from loading gantry/control room/weigh bridge exists?
e. Whether a water sprinkler system adequately covers tank of the truck?
f. Whether provision to stop pump from loading gantry/operators' cabin exists?

iv. Training:
a. Whether pocket-size instruction booklets/manuals in widely understood languages have been issued to all concerned?
b. How it is ensured that loading operations such as hose connections, etc., are always done by plant operators rather than tank truck crews?
c. Whether operating procedures have been displayed in the vicinity of tank truck loading area?
d. Whether protective gloves are being used by the operators?
e. Whether vehicle crews have been trained on the hazardous nature of LPG and to handle emergency situations?

VII. Captive power plant, electrical substations and utilities:

i. Special requirements:

Audit is done as per OISD-145-9 as below:

a. Whether single line diagrams of electrical power distribution are updated and as built?
b. Whether the installation is duly approved by the Chief Electrical Inspectorate to the government for energisation (DGMS for oil field installation)?
c. Is there any unauthorized installation in operation, including temporary construction power at the installation?
d. Whether actions were taken on IE inspector's visit report?
e. Whether electrical installations/repairs to existing installations are being carried out through licensed electrical contractors under supervision of licensed electrical supervisor (approved by state govt.) as I.E. rule no. 45?
f. Are the electrical equipment installed conforming to the area classification (IS: 5572) approved by CCE?
g. Whether there has been revision in the hazardous area classification from the original (due to modifications/expansions, etc.) and the electrical equipment was accordingly replaced to conform to the revised classification?
h. Whether the suggested remedial measures given in the accident investigation reports in last 5 years? Report, if any, have been implemented?
i. Whether the list of electrical authorized persons is being maintained (I.E. rule no. 3) and displayed?
j. Whether inspection of electrical equipment is carried out as per approved schedule? If, yes, show inspection reports.
k. Whether an approved work permit system is being followed?
l. Are the installations generally rust/corrosion protected effectively? (such as using cadmium plated bolts/nuts, painting at regular intervals, etc.).
m. Whether the substations/power plant rooms/control rooms/MCC rooms, etc., are used as storage areas for scraps, oil drums, clothes, spares and such foreign materials?
n. Are the flameproof features maintained intact for flameproof equipment? (any damage or missing bolts/nuts in the junction boxes/panel covers/terminal boxes, etc., convert the equipment to non flameproof).

o. Whether flameproof telephones are used in the classified hazardous areas?

p. Whether all transformers and switchyards are free from vegetation/dry grass?

q. Whether all the electrical equipment (switch boards, relays, motors, transformers and bus ducts, etc.,) and junction/terminal boxes, etc., are vermin proof?

r. Whether preventive maintenance schedules as per OISD-137 are carried out?

s. Whether CO_2 fire extinguishers are provided for electrical fires?

t. Whether sand-buckets are provided?

i. Substation (Equipment 7 Building):

a. Is the substation equipment freely accessible for operation, inspection and maintenance?

b. Are the minimum clearances around the equipment maintained as per I.E. rule 51?

c. Are all the substation equipment provided with double earthling connections?

d. Are the live parts of the equipment made inaccessible from inadvertent contact by barriers/shrouds)?

e. Whether the rubber gloves and mats with voltage test seals are provided in front of the switchboards and are in good condition?

f. Whether switchboards are dust/vermin proof? Is pest control treatment given for substation?

g. Are unused cables entries left open?

h. Have the protective relays been set at the recommended values (as per design)?

i. Whether the fuses used are of HRC type and of ratings specified in the single line diagrams?

j. Whether circuit identification marks/tags are provided or not?

k. Are the relays and meters tested for proper functioning/calibration as per approved schedule?

l. Whether the bus bars are provided with distinct color codification (refer yellow, blue and black)?

m. Whether the conduits used for cable entry are sealed and earthed?

n. Are the energized electrical equipment provided with caution notice?
o. Is the substation building made water tight and all the openings in walls (such as cable trench entry/bus duct) sealed?
p. Whether any water stagnation is there inside the cable trenches?
q. Are all the power (HT & LT) and control cables in trenches properly segregated?
r. Are the cables near the terminations clamped?
s. Is there any combustible material used for the construction of the substation building?
t. Whether the cable trenches are provided with concrete slabs/checkered plates?
u. Are the earth electrodes periodically tested and maintained properly?
v. Are the earth resistances of the grid measured, checked and recorded periodically?
w. Is the substation provided with emergency lighting?
x. Is the substation provided with a telephone?
y. Is the trench sump pump (for de-watering) in working condition?
z. Are floors kept clean?
aa. Whether switch gear rooms located in the hazardous area are pressurized? If not, are these provided with flameproof equipment?
ab. Whether all panel doors are kept closed?
ac. Whether all the indicating lamps in the panel are in working condition?
ad. Whether oil immersed starters filled with oil to the required level and of dielectric strength?
ae. Whether a shock restoration chart and first aid box is provided?

ii. Switchyard and outdoor equipment:

a. Are the following components individually earthed with two separate earth leads (No. 8 SWG copper or equivalent)?
    - metal frameworks,
    - A.B. Switches and circuit breakers,
    - D/O and H/G fuses,

- poles/switches/insulator pins/stay clamps,
- operating handle of A.B. switch.

b. Whether the vertical and horizontal clearances from O.H. conductors are maintained as per I.E. rule no. 79?

c. Are the earthed electrodes provided in sufficient number? Whether the earth pins are properly maintained and test carried out as per schedule?

d. Whether the lightening arrestors are connected after the A.B. switch, for the protection of transformer/circuit breaker?

e. Whether a spate earth electrode of proper size is provided properly for the lightening arrestors?

f. Is there O.H. shielding for lightning protection and whether it is properly maintained?

g. Whether there is locking arrangement for the transformer bay to prevent entry to the yard by unauthorized person? If so, is it being followed?

h. Whether the doors of the gate are rusted and need to be painted?

iii. Cable network/terminations and joints:

a. Have the route markers and joint location indicators been provided in a permanent way throughout the plant and offsite areas?

b. Whether the underground cables in excavated trenches are laid at designated depth with sand filling, brick protection? Check (on random trial pit basis) is underground cable exposed at any location?

c. Are the cables provided with identification tags at intervals of every 30 meters? (on random trial pit basis).

d. Whether all the cables throughout their route are protected mechanically from any external damage?

e. Are the flameproof type terminations for cables carried out in the classified areas (Zone 1)?

f. Are the cable joints staggered and having identification marks (when two or more cables laid in the same trench)?

g. Is the oil seepage from process units getting drained through or collected in the cable trench/manholes?

h. Whether the armor of the cables exposed at the terminations?

i. Are the cables directly taken to the terminals without suitable cable glands/clamps/armor earthling?
j. Are the cables above ground supported with proper clamps?
k. Whether the cables (telephone/power/lighting) are taken through an overhead system? (inside refinery/process units/other areas). If so, what is the arrangement?
l. Whether one run of earthling strip is fixed through the entire length of cable tray (wherever cable tray is used)?
m. Whether proper mechanical protection and water sealing arrangements are provided to cables specially at the following locations?
    - road crossings,
    - railway installations,
    - cathodically protected u/g piping,
    - drainage and other u/g facilities,
    - RCC structure/Brick walls,
    - where cables rise above ground.
n. What are the fire protection arrangements adopted for cables laid above ground?
o. Are the G.I. pipes/conduits used for cable entries (sleeves) sealed at both ends, conduits earthed?
p. Are all conduit ends for cables sealed at both ends?

iv. Motor control stations:
  a. Whether there is annunciation facility in case of failure of air supply (wherever air purging is used to achieve flameproof quality)? Is it in working condition and monitored regularly?
  b. Is the approved inspection schedule followed?
  c. Whether terminals/junction boxes are properly sealed and covered?
  d. Is there any unauthorized modification carried out (by way of local fabrication, etc.,) on the flameproof motors /control stations/terminal boxes? (To accommodate XLPE cable termination, etc.)
  e. Whether insulation resistance values between phases and between phases and earth of the motors are within acceptable limits? (Select a few critical motors at random).
  f. Whether emergency isolation devices for the motor is installed near the motor?

g. Whether double earthling provided is from two separate distinct earth connections of the grid?
h. Are the motors (75 KW) provided with space heater and local ammeters?
i. Whether painting of motor body, control station and supporting structure are in good condition?
j. Whether there are motor guards provided to prevent inadvertent contact over the moving shaft?
k. Are the name plates of motors worn out? Whether the inscriptions are visible?
l. Whether the drive belts are antistatic fire resistance type?

v. Lighting installation:
a. Whether the separate distribution for the lighting system is adopted from the main switchboard, through sub distribution/fuse distribution and lighting panel final circuit boards?
b. Whether lighting provided is sufficient? (Compare with IES chart for required levels of illumination at selected levels).
c. Is there an auto changeover system provided from a normal to emergency system? If so, is it functioning satisfactory?
d. Whether the body of lighting fixtures is separately earthened in addition to cable armor earthling?

vi. Earthling and bonding:
a. Is the earthling carried out as per IS: 3043?
b. Is the earth leakage protection wherever provided functioning satisfactory?
c. Are the earth electrodes being tested periodically?
d. What is the grid resistance (in ohms)? Is this periodically measured and recorded?
e. Whether all 3-phases equipment double earthed and single phase equipment single earthed?
f. Are all the loading arms used inside the plant earthed?
g. Is there earth continuity between flexible hoses and loading arms?
h. Are the tanks, vessels, process piping, steel columns, metallic structure/sheds and building/fencing, etc., earthed?
i. Are the hardware used free from rust and corrosion?
j. are the earthling connections to lighting protection and

electrical system protection (natural earthling and body earthling) provided separately?

k. Whether there is any discontinuity of earthling connections?

l. Whether flexible wires used for tanker body earthling is soldered to crocodile clips?

m. Whether crocodile clips lost the sprung action causing improper grip with the surface?

n. Whether earth jumpers across the pipe flanges are provided?

o. Is the buried underground earth conductor getting exposed at any location?

p. Whether the earth pits with chambers are exposed and visible or is embedded inside the loose earth?

q. Whether the earth strip connections on the earth electrodes are loose/sheared off or tight?

r. Are the weigh bridge platforms earthed?

s. Whether the earth strip joints are having an overlapping equivalent to the width of the strip (minimum) and all four sides are welded?

t. Whether all products and utility pipelines are earthed at every 25 meters apart and also at the point where it enters and leaves a shed where volatile liquids and vapors are handled?

u. Whether the earth strip runs continuously through the conveyers and supporting structures also earthed?

v. Whether all electronic equipment which stores electrical energy are properly earthed?

vii. Captive generation:

a. Is the neutral of the generator provided with an isolator?

b. Whether the neutral of the generator is earthed by not less than two separate and distinct earth connections and whether the earthling is after the isolator?

c. In case of high voltage generators (voltage above 650 V) whether the neutral point is earthed by two separate and distinct connections with earth each having its own electrode?

d. Whether the exposed wires near the generators/generator to switch board is in metallic conduit and the conduit earthed?

e. Whether the control panel for the generator is earthed with two separate and distinct connections with earth?

f. Whether generator panels are made vermin proof?

g. Whether more than one generator is provided, whether provision of an individual breaker is kept for parallelizing arrangement?

h. Whether a danger notice provided for the generator enclosure?

i. Whether the location (where generator is installed) is dry and dustfree and nearer to the loads?

j. Whether provided with fire and weatherproof enclosure with ventilation?

k. Whether exhaust provision is far away from the enclosure?

l. Whether control panel is available near to the generator equipped with all protections (short circuit, over current, earth fault etc.)?

m. If synchronized with grid, whether effective measures are available to prevent the back feeding?

n. For prime mover whether the fuels like Naphtha, LSHS are used, additional safety precautions taken?

viii. Battery/charger/UPS:

a. What is the type of battery used?

b. Whether the battery room is clean and well ventilated? Are the floors acid proof?

c. Whether exhaust fans provided?

d. Are the cell voltage, specific gravity of cell, and level of the electrolyte being checked and recorded as per approved schedules?

e. Is there any corrosion in terminals/connectors?

f. Whether all the safety equipment (Rubber apron, goggles, acid proof hand gloves, etc.) are available?

g. Is the operation of UPS satisfactory? (test for critical services).

h. Whether trickle charger is working and checked properly?

ix. Electrical heat tracing:

a. Whether all indication lamps are functioning properly in the panel to ensure health of tracer circuits?

b. What is the automatic control or protection provided for temperature raising above allowable limit? (Especially at hazardous area) Is this being tested periodically for proper functioning?

c. Are there damages to insulation/tracers?
d. Is the system water tight?

x. Portable apparatus/Mobile equipment (battery operated):
a. Is the electrical system of the portable apparatus/mobile equipment provided with intrinsically safe/flameproof features when required to be taken to the classified hazardous area? (Check test certificates).
b. Are safe voltages (24 Volts) being maintained for use of portable appliances such as hand lamps, torches, etc. (used for inspection/testing) especially in classified areas?
c. Whether the flexible cables used for portable tools are free from damages and not with too many joints? Are the cables protected from mechanical damages (by providing metallic sleeves)?
d. Whether the cables used for portable tools/equipment are 3 core for single phase and four core for 3 phase equipment respectively? Whether the respective $3^{rd}/4^{th}$ core is used to earth the metallic body/screen covering of the cable?
e. Are the power cables used for portable equipment such as welding machines, generators, switchboards, etc. provided with armoring and gland type terminations? Is the armoring bonded to earth?
f. Are the portable tools/equipment being checked every time before use that there is no earth leakage? Is it ensured safe to use, by a competent person before using?
g. Are the portable equipment including the electrical system in the mobile equipment/vehicles (Battery operated), and test instruments and tools inspected regularly?
h. Whether all wiring and current carrying parts of an industrial electric truck, so constructed and enclosed as to protect from mechanical damage?
i. Whether any un-insulated current carrying part is exposed specially on the outer surface of the truck?
j. Whether all the components are accessible for maintenance/repairs?
k. Whether the battery has the protection by means of an non-combustible enclosure?
l. Are the battery terminals provided with protective rubber gloves?
m. Whether the battery enclosure is provided with sufficient

ventilation to minimize the possibility of accumulation of explosive hydrogen air mixtures above the battery?

n. Are there reported accidents due to improper maintenance/ the wrong use/wrong selection, etc., of portable tools/mobile equipment (electrically operated) in the last two years? If so, verify the accident investigation reports and compliance of recommendations.

o. Whether connecting a welding machine the return path of the welding machine is directly connected to the work piece with proper insulation?

VIII. Firefighting and safety:

VIIIA. Organization and administration:

i. Management philosophy:

a. Whether the declared safety policy is made available to all concerned in written form?

b. When uploaded last?

c. Do the employees contribution towards safety form a part of the appraisal system?

ii. Procedures:

a. Whether procedures are well defined and manuals are available and accessible plant/facility wise for the following functions?

- Operation
- Mechanical.
- Inspection and Maintenance.
- Electrical.
- Instruments.
- Civil.
- Firefighting and Safety.
- Emergency and Disaster Management Plans.

b. When were these documents updated last?

iii. Safety audits records:

a. When was last external/internal safety audit carried out?

b. Whether accepted safety audit recommendations are implemented within time bound schedules?

iv. Safety committee meetings:

a. Whether safety committee meetings are held regularly?

v. Safety consciousness and propagation:

a. Whether following items have been taken care of through display of (in languages widely understood)?
- Sequence of operations where critical/batch operations are performed.
- Safety posters/slogans.
- Safety precautions at strategic points through sign boards.

b. Whether fire and safety training programs are conducted periodically and are in line with OISD-154?

c. Whether safety awareness competitions are held periodically? What are the types, frequencies, participants? Any other motivation exists?

vi. Work permit system:

a. Whether work permit system meets the requirements of OISD-105/137?

vii. Contingency/disaster plan:

a. Whether disaster management plans are available for different possible types of scenarios?

b. Whether all the concerned personnel are given their individual copies?

c. Whether these copies are available at central place, for emergency uses?

d. Whether these plans are approved by concerned authorities?

e. Are mock drills being conducted?

f. Whether the feedback generated from mock drills is utilized to improve plans?

viii. Plant security:

a. Whether the installation has boundary walls/fence as per prescribed norms?

b. Whether gates are manned?

c. Whether identity badges/cards are issued and worn by employees, visitors and contractor personnel for admission to restricted areas?

d. How it is ensured that visitors/contractor personnel do not trespass into non permitted areas?
e. Whether frisking is done for all people at gates? If so, how much percentage is covered in each category of people?
f. Whether a Crisis Management Scheme for Bomb threat exists?
g. Whether watch towers exists across the boundary walls?

VIIIB. Industrial hazard control:

i. Rating of house keeping:
a. Operating areas.
b. Workshops.
c. Stores/go down/yards.
d. Loading gantries.
e. Effluent treatment facilities.
f. Tank farms.
g. Storm water channels.
h. Utility block.
i. Electrical substations.
j. Fire water tanks.
k. Any other areas.

ii. Machine guarding:
a. Rotary equipment operating.
b. Areas.
c. Workshops.
d. Others.

iii. General safety features:
a. Unprotected floor openings.
b. Unprotected areas.
c. Slippery defective floors.
d. Stairway surfaces.
e. Inadequate illumination.
f. Ventilation.
g. Cross-overs over piping.
h. Platforms/handrails.
i. Walkways in tank farm areas.

j. All intrinsic safety devices (For refinery, ref to 1.7 of OISD-145-04)

iv. Maintenance of tools and tackles:

a. Whether all lifting tools and tackles are tested as per statutory requirements and record maintained?

v. Material handling:

a. Whether safe lifting methods and proper protective equipment are used?

vi. Personal protective equipment (PPE):

a. Whether requirements are identified as per operating manuals/standards?

b. Are these readily available at the time of need?

c. How periodically these are inspected and records maintained?

d. Whether personnel are trained for effective use?

vii. Toxic materials and occupational health:

a. Is there any toxic material in use?

b. Whether relevant PPE is available as per requirement?

c. Is storage of toxic material proper?

d. Is disposal of containers of toxic material safe?

e. Is there display of Do's and Don't's for handling of toxic chemicals?

f. How the health monitoring is done? Whether records are maintained?

g. Is First aid available for toxic chemicals handled?

h. Whether proper communication system exists?

i. Is there any monitoring of work environment for toxic chemicals/gases, etc., and records maintained thereof?

viii. Employee training:

a. Whether all the training given to all employees (operators/technicians/engineers) is in line with OISD-154 w.r.t. following items?

- Syllabi.
- Course materials.
- Class room/field raining.
- On the job training.

- Fire and safety.

b. Whether refresher courses are conducted for technical personnel?

c. What is the evaluation and qualifying system?

d. How employees are motivated for safety performance?

e. How are the contractor personnel trained/briefed w.r.t. safety procedures?

f. Whether the Tank-Truck crews are trained as per C.M.V. rule 9?

VIIIC. Fire, accident and near misses:

i. Reporting formats:

a. Whether reformats as per OISD-GDN-107 are available for the following separately?

- Fire and explosion.
- Accident causing injuries to personnel and damages to equipment and products.
- Near Misses.

ii. Reporting procedures:

a. Whether responsibilities are clearly defined as regard to preparation of reports?

b. Who receives the reports?

c. Whether all minor incidents and near misses included as reported to Safety Dept./concerned authority?

iii. Follow-up recommendations:

a. Implementation:

- Whether responsibilities are assigned for implementing recommendations?
- Whether recommendations implemented with time bound schedules?
- How the recommendations arising out of previous accidents are implemented?

b. Dissemination:

- Whether all procedures/systems are established and streamlined for dissemination of information?
- What is the MIS up to chief executive level with respect to information on fires, accidents, and near misses?

iv. Maintenance of records and data:

a. Accident statistics of last how many years available?

b. If required, how any case history last could be located?

- with respect to specific case based on particular area/type.

v. Review of accident data:

a. Is there any system to periodically review the trends of accidents for taking necessary actions?

b. Whether any review undertaken at top management/corporate level periodically to review the safety performance and new goals are set up?

vi. Fire protection:

a. Are the facilities adequate and reliable for anticipated fire?

b. Is manpower available and effectively trained to handle the firefighting equipment?

c. Are emergency control procedures adequately established?

vii. Philosophy and standards:

a. Does basic philosophy of designing the firefighting facilities follow OISD-116 (for refinery), OISD-117 for (Marketing-POL) and OISD-144 (LPG Bottling plants) as necessary?

viii. Facilities:

a. Whether facilities are provided adequately and have the necessary efficacy?

ix. Fire water:

a. Fire water storage:

- What is the capacity and how it was determined?
- What is the power source? How many are diesel and electric driven?
- What is the standby philosophy?
- What is the capacity of each pump and at what head?
- What is the priming system?
- Is there any jockey pump?
    - what is the capacity?

  - what is the set pressure?

- What is the fire water pumps' starting system-manual or auto?
- What is the condition of battery of diesel engine?
- What is the diesel fuel tank capacity in hours?
- How frequently does the jockey pump start? How long it runs?
- Are the fire water pumps/engines started and observations logged as per OISD-116/117/144?

b. Fire hydrant system:
   - Network-grid type/closed type/any other?
   - Mains?
   - What are the materials of construction of pipes?
   - How is corrosion protection is ensured?
   - How is the sectionizing of hydrant system?
   - How is the spacing between the hydrants?
   - Are the monitors/hydrants properly distributed in critical areas and all facilities covered?
   - Are these hydrants/monitors easily accessible and operable?
   - Is the layout vulnerable at any points from the hazard?

c. Foam system:
   - What are the types of foam available?
   - What is the dilution potential and quantity?
   - What are the storage arrangements and its accessibility?
   - Do the coverage and application rate for floating roof tanks/fixed roof tanks meet OISD norms?
   - Are foam branch pipes, nozzles, wheeled equipment available and adequate as per OISD standard?
   - Proportioning equipment-suitable for pick up?

d. Fire station:
   - Is the location away from hazardous area?
   - Are numbers and variety of vehicles sufficient?
   - Facilities on fire vehicle capacity, pressure, flow,

proportioning system, etc., daily starting of vehicles and other check logs?

e. Fire extinguishers:

- Are the portable extinguishers meeting OISD requirements? Are they BIS approved?
- Whether periodical inspection, maintenance and testing of fire extinguishers are promptly carried out and proper records maintained in line with OISD standard?
- Is replacement of extinguishers on shell failure or hydrotest ensured and necessary records available?
- Any quantity of cartridge and powder sufficient at work site and spares available in store in line with OISd standard?
- What is the monthly consumption of cartridges and products?

f. Fire hoses and accessories:

- Are numbers of hoses sufficient with respect to hydrant in the system?
- Are the hose boxes strategically placed?
- Are the hoses easily removable during an emergency?
- Are the hoses periodically tested/checked and replaced?
- Are the hose boxes painted lemon yellows as per OISD standard?
- Are adequate sand drums/buckets available?
- Are all various types of nozzles/foam branch pipes, etc., provided as per OISD standard?

g. Protective/safety equipment:

- Are safety appliances available in line with OISD standard?

x. Inspection and testing facilities:

a. Pumps:

- Is flow vs. pressure meeting the requirement?
- Is lubrication periodically carried out?

b. Hydrants:

- Is pressure and flow at different points adequate?
- Are all valves easy to operate?
  How effective is the fire water spray system?
- Is foam proportioning equipment in working order?
- Whether hydrotesting and regular inspection of fire water gate mains is undertaken and records maintained?
- Whether various checks and records of periodical running of diesel engines is carried out?

xi. Fire hose and accessories:

a. Are hoses, nozzles, adapters, hose holders, wrenches, etc., inspected regularly?

b. Foam concentrate:

- what is the inventory?
- Are the procurement/expiry date records maintained?
- Whether protein foams are chummed regularly and records maintained?

xii. Additional items:

a. Are the firefighting facilities (fixed as well as portable) provided at the port meeting requirement as per OISD standard-156?

xiii. Others:

a. Whether the fire siren and fire alarm are provided at strategic locations and are audible up to the desired range?

b. Whether siren codes are identified for different types of fire/emergencies?

c. Fire siren and fire alarm network (electrical/mechanical/ hand operated) are working satisfactorily?

d. Ambulance and first aid services with complete kits?

e. Contingency plan for medical evaluation, etc. and first aid training to personnel on shift?

f. Availability of emergency lights in critical areas?

g. Availability of special firefighting components, aprons/ suits/emergency flood lights, power megaphones, etc.?

VIIID. Organization procedures on firefighting:

a. Organization:
   - Whether the organogram of firefighting and other supporting services are well defined?
   - Is periodic fire drill conducted, shortcomings observed and necessary actions initiated?
   - Do locations have well defined fire order?
   - Whether organogram has provision to meet emergency situations during idle shift/Sundays/holidays?
   - Whether relevant fire order is displayed at various work stations indicating station employees role during fire/emergency?
   - Whether following documents are updated and displayed as required?
     - Organogram.
     - Fire hydrants layouts and P&ID diagram.
     - Firefighting procedures/manual.

b. Disaster management plan:
   - Whether mutual aid schemes are made with neighboring industries, local fire brigades, etc.?
   - Whether resource requirements are worked out?

c. Training:
   - Whether training provided to fire crew/other employees and CISF etc., is in line with OISD standard?

d. Checklist for DMP:
   - What is the frequency of DMP drill? When was it last conducted? Whether the shortcomings observed during the drills were analyzed and actions taken to prevent repetition?
   - Whether mutual aid schemes are planned with neighboring industries, local fire brigades, etc.? is the scheme periodically discussed with members? Are they trained to fight fire in oil/gas installation?
   - Whether the resources of the operating location and the mutual aid members are properly tabulated? Whether assessment of necessary man power and equipment was studied and action taken against shortfalls?
   - Whether organogram of various co-coordinators,

succession charts showing alternate persons and sub-organization charts are well displayed?

- Are responsibilities of various neighboring industrial units discussed and recorded? Whether the duties of various coordinators and team members are well defined?
- Is the following information included in DMP?
- Name, address, telephone numbers (official and residence) of anchorages/key persons are clearly shown for the following?
  - Mutual aid members.
  - Police, fire and district authorities.
  - Hospitals and doctors.
  - Medical shops, ambulance, blood bank, blood donors.
  - Water supply dept, water transporter, electricity board, railway, port trust, telephone exchange.
  - Generator hire service, transport hire service, welfare bodies, social organization and catering service.
  - Emergency control centre planned and properly furnished/fitted to meet the requirements.
  - DMP is based on MCA and risk analysis.

3.15.2 ESA

I. Methodology adopted to carry out the audit:

i. The audit team should consist of three to four members, with one member heading the team. An audit generally takes three to four days to conclude.

ii. Team should conduct an opening meeting with all concerned relating to the installation where the audit is to be conducted. Audit agenda is clearly demonstrated in the meeting.

iii. The audit team members divide the responsibilities among themselves and each member takes care to conduct the audit as per planning.

iv. From the installation, there should be a particular coordinator engaged with respective ESA team members.

v. As a part of the audit, the following should be taken care of:

- Study of operation manuals, operation and maintenance log sheets, Inspection documents, quality control documents,

PandID, and layouts of the units, emergency preparedness plan, Disaster Management Plan (DMP), etc.

- Study of the fire and safety manuals and safety management system manual.
- Checking the incident/accident reports.
- Review of ISA audit report on coverage, findings, recommendations and action status.
- Interview of personnel involved in operation and maintenance of the process units, off sites and utilities, store keepers in materials department and technicians in workshop and garage.
- Physical condition tour of the industry from process units to off sites—including effluent treatment plant, firefighting facilities and storage and dispatch—up to power and utilities units.

vi. The ESA team should plan for a major fire mock drill one day and evaluate the firefighting system effectiveness in line with OISD standard.

vii. The ESA team conducts a closing meeting on the last day of the audit, where the good aspects of the safety system and practices by the audited installation are highlighted and appreciated; next, the shortcomings are deliberated and discussed in detail. There should be a scope for debate in the meeting. The action points are reconciled in the meeting itself with the participants from the audited installation and the ESA team advise the participants to make an action plan to liquidate the recommendations and submit the draft recommendations to the audited industry/installation.

viii. The ESA team after reaching their office, submit the report to concerned authorities and authorities after review of the issues prepare the final report and circulate to all concerned, including the audited installation.

II. Content of the report:

The report should contain the following:

i. Observations and recommendations:

ii. Detailed findings as per stipulated guidelines, e.g., for petroleum refineries, as per OISD 145 guidelines.

iii. Physical condition checks.

iv. Findings and recommendations on major fire mock drill.

Observation and recommendation as in sl. n. (i) should contain the following:

- Noteworthy observations.
- Observations which call for improvement.

In detailed findings, all detailed observations should be listed for each process unitwise, for each subsection of offsite; (i.e. tank farm, effluent treatment plant, and so on); each dispatch section (i.e. tank truck dispatch, tank wagon dispatch, pipeline dispatch, if any, pressure vessel dispatch, etc).

The observations should be documented in proper format and in simple English for the purpose of good communication. There should be three columns in the format; namely sl. no., description of the item/area audited and the auditors' observations.

3.16 Safety Culture

With full application of all knowledge and rules, safety in the plant cannot be ensured on a sustained basis because safety is a subject of habit to be inculcated in the system of plant operation/organization concerned.

Following are the elements of safety culture to be followed on a day-to-day basis within the organization to keep the operation sustained, stable and effective:

I. Wearing personal protective equipment (PPE) should be religiously followed and the system should be enforced; a display board for hard hat areas should be properly placed and people must use helmets in such areas.

II. All fire extinguishers, and fire hoses should be available in respective areas, as per document and approval.

III. Fire tenders and all equipment and control systems in the fire and safety department should be in good health and operating condition.

IV. Fire and Safety Department should be able to work in liaison with the state firefighting station and district administration in case of emergency.

V. All necessary documents for operation and maintenance should be easily traceable.

VI. Pocket operating manual for each area should be available and to be approved by plant in charge.

VII. There should be a proper charge handing/taking over system among the rotating shift employees.

VIII. Operating log books should be properly filled in and countersigned.

IX. There should be a continuous quality control system for operation and it should be reflected in signed documents.

X. Planned maintenance and inspection should be followed and should be reflected in the records.

XI. There should be an everyday operation and maintenance review meeting at site for a short duration where all multidisciplinary groups like operation, maintenance, inspection, quality control, dispatch, fire and safety, power and utility and technical service representatives should be present.

XII. The above operation meeting should start with a safety talk and the fire and safety department should highlight the occurrence of near-miss, minor/major incidents, if any.

XIII. There should be a regular safety tour throughout the plant by a senior management representative, be it weekly or monthly.

XIV. Each meeting inside the organization should start with a safety talk.

XV. Apex management authority should take a review of plant operation safety and health on a regular basis, generally on a quarterly basis, at minimum.

XVI. Any change of management in facility and operation should be approved by a competent authority and that too after following HAZOP studies.

XVII. Changes management approved should be duly incorporated in the respective document.

XVIII. Emergency management systems and procedures should be established with proper documentation, hands-on training, well designed flow schemes, plants and machineries and control systems.

XIX. Training need identification is to be done for each employee and should be executed.

XX. Any employee should be entrusted with operation responsibility after undergoing a training course for a certain period and should take charge with approval of the plant in charge.

XXI. Boiler operators and firefighting operators can take charge if certified with their respective statutory authority.

XXII. Safety awareness surveys among the operating employees should be conducted at least once a year by an authorized external agency.

4

Project Safety

Project activity in any installation is a one-time job for a particular project; if the industry feels they would have project activities from time to time, they then set up a project department; alternatively, they execute the project through external agencies. In light of this, project safety has not been included in sl. no. 3.

For project execution in India, OISD STD 105 on work permit system and OISD-GDN-192 on construction safety have already set the guidelines; these should be religiously followed; additionally, SP 70 and NFPA codes provide adequate guidelines for safe execution of projects. Also, in project execution, certified OHSA (Occupational health and safety hazards) guidelines compliance is mandatory. The project planning should contain an assurance plan in line with all international standards. An EPC contractor should have a proven safety track record to be ensured.

The EPC contractor should submit the safety organogram to PMC (Project management consultant) who, on behalf of the owner organization, has to approve the same before the project is executed by the EPC contractor. The EPC contractor, additionally, has to ensure the issue of a daily safety plan by his safety officer: the engagement of a safety officer in charge in a safety organogram is a must. In project execution, at different stages of execution and at regular intervals, a management safety review is to be undertaken under the chairmanship of a competent authority in the owner organization. This review can be clubbed with the management review on job progress.

In project execution, generally, five standard formats are followed to be complied with at different stages of execution as follows:

- Format 1: A report is issued by the executor, indicating mechanical job completion with reference to P&ID numbers to be attached with marked up P&IDs and remarks. The main receiver of the report is the operation in charge. Here, mechanical completion includes hydrotest of pipelines without installation of instrumentation and with remarks on equipment installation completion status.
- Format 2: A report is issued by the operation in charge to the project executor, indicating comments on Format 1 in reference to P&IDs.
- Format 3: A report is issued by the project executor, indicating compliance with all comments as in Format 2 above; while issuing Format 3, the executor confirms the installation of equipments in all

respects, including alignment of the equipment with drive and civil structures in the referred circuit.

- Format 4: Again, this report is issued by the project executor, after the operation in charge concurs with the Format 3 report issued by project executor, with subsequent pre commissioning jobs completed by operation in charge followed by completion of all instrumentation and electrical jobs, along with process control and related offsite facilities. This report goes to a competent authority in the organization, who in turn issues commissioning clearances.
- Format 5: A report is issued by the operation in charge indicating completion of plant commissioning, along with all comments for future incorporation.

4.1 Construction Safety

Following standards are applicable in construction safety:

- SP 70 from BIS and OISD-GDN-192: Construction safety practices.
- NFPA 70-E: Standard for electrical safety in workplace, 2004 edition.
- NFPA 30: Flammable and combustible liquids code.
- NFPA 33: Standard for spray application using flammable and combustible material.
- NFPA 241: Safeguarding construction, alterations and demolition operation.
- NFPA 70: National electric code.
- NFPA 51B: Standard for fire prevention during welding, cutting and other hot work.

In India, the construction industry is the second largest employer next to agriculture; whereas in terms of accidents it is second highest next to road accidents. The number of fatalities occurring from construction work in India is quite high with maximum incidents like falls from a height and in confined places. The mobile nature of the workforce in construction poses a great challenge in ensuring that all of them are adequately trained. Construction safety management indeed is a great challenge due to the involvement of an unskilled, illiterate and mobile workforce. Since the projects are located in remote areas in many cases, the surrounding population involved in construction is substantial. These personnel are generally from an agricultural background, speaking and understanding local languages only. This poses an additional challenge due to limited communication.

Hence, unique safety programs and mechanisms should be developed to overcome this. With strong planning, effective implementation and continual training with focused safety management, a good safety record can be achieved comparable to the international level.

The average fatal accident rate (FAFR) in Nuclear Power Corporation (NPCIL) for 5 years was found to be 0.22 incidents/1000 employees/year as against an estimated value of 15.8 for Indian construction industries. Also, the above fatality figure of NPCIL was well comparable with industries in the United States for the year 2005 which was 0.23 (Ref; published by US Dept of Labor in 2005, NPCIL data in Dec., 2007).

To enhance safety in construction work, we should place emphasis on the following aspects:

- Innovation in training methodologies to achieve a higher effectiveness of training among the contractor employees. Training capsules from internationally professional institutes should be used. Also, training capsules from NPCIL (having very good record of safety data) can be used.
- Developing and implementing a behavioral-based safety program to improve orientation of the workforce towards safety in work.
- Implementation of innovative engineering measures to strengthen the safety requirements at design stages to achieve a safe working environment during construction. In this regard, a benchmarking study should be conducted on safety among different technology and engineering houses.

4.2 Training

Required training and certification of managers and others responsible for construction activity in industrial safety are essential to enhance their perception and appreciation for industrial safety.

The role of the line managers and safety professionals in preventing unsafe incidents is quite important. Therefore, it is necessary that safety requirements are assured on a regular basis by scrupulous field rounds and the deficiencies identified are attended promptly.

Standard training modules have to be evolved as discussed above. Use of modern pedagogical training aids such audio-visuals and in-field training are effective to enhance safety in construction. Over and above, feedback mechanisms through direct communication by line functions strengthen the effectiveness of the training.

4.3 Safety Management in Construction

The attributes and requirements to achieve effective management of safety right from the design stage to execution and operation must be identified and addressed appropriately through a structured program. To achieve this objective, it is imperative to recognize the important elements of process safety management as mentioned earlier and strengthen the same at each stage.

For a construction project, the most important elements which are to be considered in a management system are as follows:

4.3.1 Safety Organogram

A well-designed safety organization for contractors and sub-contractors and interface with the department is essential. Implementation of safety is a line management function, thus ownership lies with them; however, line managers should be backed up by a competent authority in industrial safety that provides expertise and supervision of the work environment and equipment, such as lifting tools, tackles, scaffolding, ladders, cranes, hydra, etc., used in construction. Scrupulous implementation and adherence to industrial safety procedured and requirements is needed to be observed at all levels as an ongoing program. Some of these systems to identify areas of improvement and achieve enhanced industrial safety status are enumerated as follows:

- Safety surveillance and safety-related deficiency management system.
- Area wise task force for enforcing safety in construction.
- Contractors' safety surveillance and correction program.
- Entry passes to work sites only after induction safety training, KYC verification and so forth.
- Periodic safety audits.

Regular communication should be ensured between the safety officer of the contractor and safety organization of the construction project. The contractor safety officer should be under administrative control of the head of safety of the organization.

The organization must develop and institute procedures, work plans and programs that are implemented with a common understanding of utility and contractor team. The regulatory requirements need to be understood and implemented in clear and unmistakable terms by all concerned, including the contractor organization.

4.3.2 Hazard Identification and Analysis to Finalize Work Procedure

This concept has been explained earlier in the relevant section above. However, to emphasize the subject in a construction project, it is worthwhile to mention that here, the activities, though planned, are carried out by a workforce which is skilled in execution of work but lack awareness of safety requirements, exhibit overconfidence and complacency at times. Hence, a regular monitoring and surveillance program, along with coaching and mentoring of employees during execution, becomes necessary to correct the aberration in safety implementation.

4.3.3 Safety-Related Deficiency Management

Even with all things in place, while the construction work is in progress, safety-related deficiencies (SRD) emerge either due to change in status in workfloor conditions or in multiple agencies working parallel. SRD also get generated with a decline in safety culture. So, it is required to detect the possible SRDs and correct them beforehand.

Following are the control measures in SRD:

- *Use of near misses and experiences feedbacks*:

  It is seen from histories of various accidents that before a serious or fatal accident we get enough opportunities to correct the unsafe conditions or unsafe practices from the minor incidents or near misses which occur as a precursor. These need to be recorded, reported and analyzed to prevent future recurrences. It is a rule of thumb in statistics that there is a major accident in 500 near misses and there is a fatality in 10 major accidents.
- *Safety meetings*:

  In order to maintain proper coordination and communication on safety aspects on a periodic basis, it is necessary to exchange regular views and experiences, as given below:
- Daily interaction between the contractor safety officer and the departmental safety in charge.
- Monthly safety meeting by each work manager of the contractor, along with the safety officer and departmental safety group.
- Sectional safety meeting for the departmental and contractor employees.
- Quarterly project level Apex Safety Committee meeting.
- Regular feedback mechanism among various agencies.
- Safety enforcement by line managers:

In order to achieve a practical solution and achieve involvement in accident prevention, safety has to be integrated with line functions. Accordingly, line managers should supervise and enforce safety requirements in the work. It is the line functions who know the hazard as soon as it is created. They have the power and resources to take immediate corrective actions. Safety personnel should act as a catalyst to enable the line managers to timely remove these hazards and any deficiency in a proactive manner.

4.4 Other Key Drivers in Project Safety

There are other key drivers to prevent accidents, which are as follows:

- Certification of line managers in industrial safety.
- Industrial safety clauses in contract conditions for effective implementation.
- Ensuring administrative control of construction activities through a work permit system, job at height passes, job at confined place, excavation permit and job at excavated place, hydra/crane movement passes and so forth.
- Encouraging mock exercises by performing models and mockup for complex work.
- Field surveillance through a structured checklist and prompt addressing of deficiencies.
- Development of a pool of line managers having industrial diplomas as a long term measure that will further strengthen the system.

5

Safety Performance Indicators

- Leading and lagging indicators

  Lagging indicators measure a company's incidents in the form of past accident statistics; for example, injury frequency and severity, lost workdays and worker's compensation costs, whereas leading indicators are the measure of proactive actions/program to prevent accidents, such as a Training & Development program for the workforce engaged, mock drills, safety audits, safety review meetings, establishing a safety culture and behavioural safety, documenting near misses and accident histories and using lessons to implement, reduction of MSD risk factors and so on.

- Why use lagging indicators?

  Lagging indicators are the traditional safety metrics used to indicate progress toward compliance with safety rules. These are the bottom-line numbers that evaluate the overall effectiveness of safety at your facility.

- The drawbacks of lagging indicators.

  The drawback to using only lagging indicators of safety performance is that they tell you how badly, but not how well, your company is doing.

- Why use leading indicators?

  Leading indicators are focused on future safety performance and continuous improvement. These measures are proactive in nature and report what employees are doing on a regular basis to prevent injuries.

- Best practices for using leading indicators:

  Companies dedicated to safety excellence are shifting their focus to using leading indicators to drive continuous improvement. Leading indicators allow you to see small improvement in performance and measure the positive ones. Best practices to improve leading indicators are as below:

  - Be aware of what people are doing vs. failing to do.
  - Be credible to performers, be predictive.

- Increase constructive problem solving around safety.
- Make it clear what needs to be done to get better.
- Track impact vs. intention.
- Keep in mind that there is no perfect or 'one size fits all', measure for safety.

- Performance scoring mechanism:
    - Scale of performance with respect to each lead and lag indicator tare to be assigned.
    - Levels of failure in lag indicators are to be assigned.
    - Levels of achievement in lead indicators are to be assigned.
    - Scoring of the organization is then to be evaluated.
    - The authorized agency is to be lined up to co-relate industries to compare and accordingly develop a benchmark.
    - Scoring is to be compared with the benchmark.

6

Safety Reporting

Requirement for the preparation of safety reports as per Indian government rules:

- An occupier carrying on industrial activity as defined under MSIHC Rules of India, 1989 is required to prepare a safety report, when the activity involves a hazardous chemical listed in column 2 of Schedule 3 of MSIHC Rules, 1989 and whose quantity is equal to or more than the quantity specified in column 4 of Schedule 3 or an isolated storage containing a chemical listed in column 2 of Schedule 2 of MSIHC Rules and whose quantity is equal to or more than that in column 4 of Schedule 2.

In general, the report should be updated regularly every three to five years.

- Purpose:
  i. To present the entire safety system to the appropriate authority.
  ii. Authorities would get an opportunity
    - to check extent of adherence to safety standards,
    - to carry out specific inspections,

  iii. To establish contingency plans, and
  iv. To take proper corrective action.
- Objectives:
  - To identify the nature and scale of the use of hazardous substances in the installation or in the isolated storage.
  - To give an account of arrangement of safe operation.
  - To give an account of arrangement for control of serious deviations that could lead to a major accident.
  - To give an account of the arrangement for emergency procedures at the site.
  - To identify the type, relative likelihood and consequences of a major accident.
  - To demonstrate that all the major hazard potentials have been identified and appropriate control measures have been provided.

- Content (in line with the requirement of Schedule 8 of MSIHC Rules, 1989):
  - The name and addresses of the person furnishing the information.
  - Description of industrial activity which includes
    - Site plans,
    - Construction designs,
    - Protection zones, explosion protection, separation distances,
    - Accessibility of plant,
    - Maximum number of persons working on the site and particularly of those persons exposed to hazards,
    - Description of the process, namely
      - Technical purpose of industrial activity,
      - Basic principles of technological process,
      - Process and safety-related data for the individual process stages,
      - Process description,
      - Operational storage,
      - Safety related type of utilities,
      - Description of hazardous chemicals, namely
        - Chemicals (quantities, substance data, safety related data, toxicological data and threshold values),
        - The situation where the chemical may occur an accident or into which they may be transformed to an hazardous chemical in the event of abnormal conditions,
        - The degree of purity of the hazardous chemical,
      - Types of possible accident(s) and hazard(s),
      - System elements or events that can lead to major accidents,
      - Safety-related components,
      - Description of safety-related units, among others,
      - Special design criteria,
      - Controls and alarms,
      - Special relief systems,
      - Collecting tanks/dump tanks,
      - Sprinkler systems, and

- Firefighting, etc.
- Information on the hazard assessment like identification of hazard(s) and risk analysis.
- Safety systems,
- Known accident history,
- Maintenance and inspection schedules,
- Guidelines for training of personnel,
- Allocation and delegation of responsibility for plant safety,
- Implementation of safety procedures,
- Implementation of risk mitigation measures like fire brigades, alarm systems and the like.
- Emergency plans containing a system of organization used to fight the emergency, the alarm and communication routes, guidelines for fighting the emergency, information about hazardous chemicals and examples of possible accident consequences,
- Co-ordination with the district emergency authority and its offsite emergency plan,
- Notification of the nature and scope of the hazard in the event of an accident,
- Action plan in the event of a release of a hazardous chemical.

- Updating of safety reports:
  - The safety report may need to be updated, when:
    - There are major changes in the plant or process,
    - New information about the hazardous chemical comes to light.

Note: In general, the report should be updated on regular basis every three to four years.

7

IFC Guidelines on EHS – Present Perspective

International Finance Corporation (IFC) is a World Bank Group that has institutionalized EHS standards as a benchmark to ensure good practices in the oil and gas industries. EHS guidelines as set by them are technical reference documents with general and industry specific examples of good industry practice (GIP). For industries in this sector, EHS guidelines are designed to be used together with the General EHS Guidelines document, which provides guidance to users on common EHS issues potentially applicable to all industry sectors. For complex projects, use of multiple industry-sector guidelines may be necessary.

The EHS guidelines give direction, highlight expected performance levels and measures from the industries. Application of EHS guidelines to existing facilities may involve the establishment of site-specific targets, with an appropriate timeline for achieving them. The applicability of EHS guidelines should be tailored to the hazards and risk established for each project on the basis of the results of an environmental assessment in which site-specific variables, such as host country context, assimilative capacity of the environment, and the project factors, are taken into account. The applicability of specific technical recommendations should be based on the professional opinion of qualified and experienced persons. When the host country regulations differ from the levels and the measures presented in EHS guidelines, projects are expected to achieve whichever is more stringent. In view of specific project circumstances, a full and detailed justification for any proposed alternative is needed as part of a site-specific environmental assessment. This justification should demonstrate that the choice of any alternative performance level is protective of human health and the environment.

Applicability in Petroleum Refineries:

The EHS guidelines for petroleum refineries cover processing operations from crude oil to finished liquid products, including LPG, propane, propylene, naphtha, motor gasoline, kerosene, diesel oil, fuel oil, Bitumen/ Asphalt, sulfur and all intermediate products being generated in petroleum refineries, including storage and dispatch facilities. There are following three sections in the guideline document, along with annexure with a general description of industrial activities. The three sections are as follows:

Section 7.1- Industry-specific impacts and management.
Section 7.2- Performance indicators and monitoring.
Section 7.3- References and additional sources.

Section 7.1 only has been discussed in this chapter as performance indicators and monitoring are dynamic and change with time and also flexibilities accepted as per respective countries' statutory obligations; however, those should be at par with the international level.

7.1 Industry-Specific Impacts and Management

Recommendations for the management of EHS issues common to most large industrial facilities during the construction and decommissioning phases are provided in the General EHS Guidelines summarized as follows.

7.1.1 Environmental

Potential issues in petroleum refining include the following:

- Air emissions
- Waste water
- Hazardous materials
- Wastes
- Noise

7.1.1.1 Air Emissions

Global warming is a most important concern being discussed at apex levels internationally in various forums. While the transport sector contributes most in this direction, industries are also not far behind them. To remain in the context of petroleum refining, furnace stacks installed at various process plants along with that of captive power generation plants of the refineries emit a lot of pollutants, of which carbon di-oxide is of major concentration, with some presence of carbon monoxide, nitrogen oxides, sulphur di-oxide and the like.

To reduce sulphur dioxide emissions, guidelines recommend installing a sulphur recovery unit inside the refinery which, refineries across all developing countries are following; on top of that to restrict sulphur dioxide emissions from furnace stacks of the process plant, guidelines recommend to use low sulphur fuel by limiting the amount of sulphur di-oxide emission to the atmosphere; the practice being followed by all developing nations to reduce greenhouse emissions. Regarding reduction of emission of nitrogen oxides, the guidelines recommend the use of low NO_X burners. The guidelines also recommend installing a stack emission monitoring system to enable assessment of emissions. In current times, many advanced instrumentations are in place to have a robust data inventory on emission control. Hence, technology selection processes for all facilities, as outlined in Chapter 3.1, play a very crucial role with respect to air emission. Guidelines also recommend carrying out hazard assessment on air emission by the refineries, as discussed in Chapter 3.2.

Regarding emissions from stacks of power plants, recommendations given with an example of a typical power plant capacity of 50 MW. It is worth mentioning that the recommendations can't be met with coal fired boilers; hence, refineries keep oil fired boilers within the premise; however, even that is also restricted with stringent norms of emissions; that is why most of the refineries are operating gas turbines on a neat basis or in combination with oil-fired boilers, which is also economical to them if a natural gas resource is available.

Regarding venting and flaring, guidelines recommend to avoid open venting, unlike in earlier days, from certain offsite equipment like LPG bullets or Horton spheres. But presently, refinery flare capacities are matched with closed flaring from these offsite storage facilities. To further restrict emissions from flaring, guidelines recommend taking alternative measures to reduce flaring; in this direction, refineries at present implement a flare recovery system; and investment on this account justifies with respect to saving of energy. Refinery flare system design and operation are presently being looked into with the following consideration, as given in the guidelines.

- Uses of efficient flare tips and flare size optimization.
- Maximizing flare gas combustion efficiency.
- Minimizing flare from purges and pilots without compromising safety.
- Use of reliable pilot igniting system.
- Installation of high integrity instrument system.
- Installation of knockout drum to prevent condensate emission.
- Minimizing liquid carryover and entrainment in the gas.
- Flare monitoring system to avoid flare lit off.
- Operating flare to control odor and to maintain no visible smoke.
- Locating flare to a safe distance.
- Implementation of flare tip maintenance and replacement program for ignitor system.
- Metering flare gas.

7.1.1.1.1 Fugitive emissions

Fugitive emissions are small amounts of hydrocarbon emissions occurring from valve glands, connecting flanges, passing valves of emergency vents, if any, passing valves in loading/unloading trucks, wagons, passing drain valves, volatile organic compounds (VOC) from ETP, storage tanks and so forth.

Guidelines to prevent and control fugitive emissions should include the following:

- Review PFDs and P&IDs accordingly.
- Select the valves, flanges, seals, packing accordingly.
- Implement a hydrocarbon recovery system to avoid open venting.
- Wherever applicable, use a gas scrubber.
- The incinerator should be at high temperature—about 800°C to ensure complete combustion of gases.
- Emission from hydrofluoric acid (HF) alkylation plant vents should be collected and neutralized for HF removal in a scrubber before being sent to flare.
- A proper prevention system should be implemented to arrest fugitive emissions from the tank farm.

In the checklisting, as discussed in Chapter 3.2, these aspects are covered in PFDs and P&IDs provided by the process licensors. The design basis is given by the refineries with respect to complete combustion of hydrocarbons, provision of flare recovery systems, providing advanced safety systems in flare control such as a good hydraulic seal, good burners, an igniting system, automated local panel with monitoring from a remote operator room, knockout drum with automatic pumping out to take care of emergencies like liquid carry over to flare system, valves/fitting with higher leakage proof rating, etc. Refineries have a sour gas incinerator system that is required to be installed at a sulphur recovery unit where the design takes care of complete burning of the sour gas. Moreover, in current days, a number of process technologies in recycling/recovery of this sour gas have come up; refineries are implementing the same on efficiency vis a vis a cost economy point of view.

7.1.1.1.2 Particulate matters (PM)

These are associated with flue gases from furnaces, with catalyst fines in fluid catalytic cracking units (FCC), other catalyst-based processes, involving disposal of waste catalysts, a coke handling system, and so on. The particulates may contain metals like vanadium, nickel and the like.

Guidelines recommend implementing the following by the refineries:

- Install cyclones, electrostatic precipitators, bag filters and/or wet scrubbers to reduce emissions. A combination of these techniques may achieve >99% abatement of particulate emissions.
- Implement particulate emission reduction techniques in coke handling as follows:
- Store coke in bulk under enclosed shelters.
- Keep coke constantly wet.
- Cut coke in a crusher and convey it to intermediate storage silo (hydro bins).
- Spray the coke with fine layer of oils to stick the dust fines to the coke.
- Use covered and conveyor belts with extraction system to maintain negative pressure.
- Use aspiration systems to extract and collect coke dust.
- The silos should be fitted with exit air filters and collected fines should be recycled to storage.

7.1.1.1.3 Greenhouse gas (GHG) emissions

Guidelines recommend reducing energy consumption to benchmark levels as outlined below:

- Operators should aim to maximize energy efficiency with respect to operation, as well as design facilities. The overall objective should be to reduce air emissions and evaluate cost-effective options for reducing emissions that are technically feasible. Additional recommendations with respect to control GHG emissions are provided in General EHS guidelines.

It is unfortunate that the energy efficiency benchmark provided by IFC more than decades ago are not being met even in the present by the refineries. As per April 30, 2007 publication of IFC, the energy consumption benchmark vis-a-vis typical actual performance of many refineries is given in Table 7.1.

As seen from Table 7.1, except land usage, actual performances in all parameters are lagging behind the benchmark; in particular, water consumption is much higher than the benchmark, while there are adequate technologies with which refineries should avail themselves to reduce water consumption. Also, refineries should improve power consumption to maximize gas turbine operation with a heat recovery steam generator (HRSG) system. Selection/modification of refinery configuration or implementing an optimization program would improve energy consumption to meet the benchmark.

It is not only the Indian refineries, but also many of the international refineries are lagging behind the above benchmark.

TABLE 7.1 Resource and energy consumption

Parameter	Unit	Industry benchmark	Actual
Land use[1]	Hectares	200-500	300-700
Total energy[1]	MJ/MT of processed crude oil	2100-2900	3300-3700
Electric power[(1)(2)]	KWh/MT of processed crude oil	25-48	55-60
Fresh Make up water	M^3/MT of processed crude oil	0.07-0.14	o.5-0.55

Note: 1. Brown field facilities. 2. Green field facilities.

7.1.1.2 Waste Water

The largest volumes of effluent include sour water containing H_2S, NH_3, phenol and other organic compounds like cyanide (in case of operation Coker and FCC unit) and so on. The sour waste water is generally processed in sour water stripper to strip out H_2S before routing the water to ETP. Also, another major effluent includes desalter water, usually called Brine; the effluents sometimes may contain entrained oil and are routed to ETP for treatment of the waste water and to recover the oils to meet the final treated effluent quality, as per the benchmark. Process waste water may also be alkaline coming from water treatment plants meant for production of DM water. In this case, waste water is neutralised separately before routing the water to ETP. Liquid effluent may also result from accidental release or leaks of small quantities of hydrocarbon products from process equipment/ plants.

Guidelines recommend implementing a management system to reduce generation of waste water:

- Prevention and control of accidental release of liquid hydrocarbons/products following regular inspection, maintenance of process, equipment, offsite and dispatch facilities.
- Provision of sufficient process fluids let down capacity to maximize recovery into the process and avoid massive discharge into oily water drainage systems.
- Design and construction of a waste water and hazardous material storage containment basin with impervious surfaces to prevent infiltration of contaminated water into soil and an underground water bed.
- Segregation of process waste water from storm water and segregation of waste water and hazardous material containment basins.
- Implementation of good housekeeping practices including conducting product transfer activities over paved areas and prompt collection of small spills.

Specific provisions to be considered for management of waste water streams include the following:

- Spent caustic from sweetening plant routed to ETP after following caustic oxidation.
- Recovery/use management of spent caustic liquor from caustic sweetening plant.
- Install a closed drain system to collect and recover leakages and spills like MTBE, ETBE and TAME and avoid sending these to ETP.
- Cool the blow down from the steam generation plant, any steam boiler prior to discharge. The effluent from the cooling tower may contains chemicals and may require treatment before routing to ETP or require additional treatment in ETP.
- Hydrocarbons in waste water to be recovered in ETP.

Installing a smart ETP for which technologies are also available gives refineries not only mileage in uninterrupted operation but also boosts morale among the employees by not affecting the livelihood of surrounding colonies, farmers and other stakeholders. There are a lot of well-established technologies for treatment of waste water in a dedicated area called an effluent treatment plant (ETP), with various sections like grease traps, skimmers, dissolved air floatation, oil water separators for separation of floatable solids, filtration for separation of filterable solids, flow and load equalization, sedimentation of suspended solids using clarifiers, biological treatment (typically aerobic) for reduction of soluble organic matter to maintain standard value of BOD (biological oxygen demand) of treated effluent water. Similarly, treatment should take care to maintain the standard value of COD (chemical oxygen demand) of the treated effluent water. Dewatering and disposal of residuals should be in designated hazardous wasteland fills. Additional engineering controls may be required for containment of VOC using a closed evacuation and recovery system, metal removal using membrane systems or other technologies, removal of recalcitrant and non biodegradable COD using activated carbon or other chemical oxidation, reduction in effluent toxicity using technologies like reverse osmosis, ion exchange and activated carbon and containment along with utilization of nuisance odours.

7.1.1.2.1 Other waste water streams and water consumption

Hydro testing water used for pressure testing equipment, pipelines and so forth should be of the same source and chemical additives are often added to the water to prevent internal corrosion of the equipment under testing. Guidelines for following pollution prevention measures include:

- Using same water.
- Reducing the need for corrosion inhibitor and other chemicals.
- If chemicals are used, they should be of the lowest toxicity, higher biodegradability, good bioavailability and bio accumulation potential.
- Well-planned hydro test water disposal management to avoid contamination if routed to the sea or surface water.

7.1.1.3 Hazardous Materials

Spent catalysts may contain molybdenum, vanadium, nickel, cobalt, platinum, palladium, iron, copper, silica and/or alumina.

Guidelines to control the emission of these metals include the following:

- Use long life catalysts and include a regeneration system to extend catalyst life cycle.
- Use appropriate on-site storage of spent catalysts and return to the manufacturer for regeneration of catalyst/recovery of metals at their end.

Present perspective: Following checklisting, technology selection and engineering standards as covered in chapter 3, it should follow the above recommendations.

7.1.1.4 Other Hazardous Waste

Petroleum refineries may contain additional hazardous wastes generated sometimes in small quantities for example, special solvents, filters, mineral spirits, spent amines, spent activated carbon filters, oily sludge from separators, tank bottoms and operational maintenance fluids. Process wastes should be tested and classified as hazardous or non-hazardous based on local regulatory requirements or internationally accepted approaches.

Guidelines recommend industry specific management strategies for hazardous wastes include following:

- Send oily sludge from crude oil storage tanks and desalter to delayed coker drum where applicable to recover hydrocarbons.
- Ensure excessive cracking is not conducted in a visbreaking unit to avoid formation of unstable fuel oil and sludge.
- Maximize oil recovery from sludge by applying mechanical or process mechanisms.
- Sludge treatment may include land application (Bioremediation) or solvent extraction followed by combustion of the residues and/or

use in asphalt where possible. In some cases, residues may require stabilization prior to disposal to reduce leachibility of toxic materials.

- Detailed procedures for storing, handling and disposal of non hazardous materials like not spoiling soil, harmful to human skins, etc., are also covered in General EHS guidelines.

7.1.1.5 Noise

Principle sources of noise in petroleum refineries include large rotating machines. During emergency depressurization, high noise level may be generated due to high pressure gas to flare and/or steam release to atmosphere. Also, during maintenance, some vehicles may also cause noise. Recommendations are given in General EHS guidelines.

7.1.2 Occupational Health and Safety

- The provisions outlined by IFC are similar to those of OSHA-18001 as being followed by the refineries if they are accredited to the system. The guidelines are discussed in the General EHS Guidelines.

- The guidelines recommend job safety analysis, viz., Hazard identification Studies (HAZID), Hazard and Operability Studies (HAZOP), or Quantitative Risk Assessment (QRA) which have been discussed in Chapter 3.2. Guidelines recommend implementation of a systematic and structured approach for prevention and control of physical, chemical, biological, and radiological health and safety hazards with the following considerations:
  - Process safety.
  - Oxygen-deficient atmosphere.
  - Chemical hazards.
  - Fire and explosions.

7.1.2.1 Process Safety

As per guidelines, process safety management includes the following actions:

- Physical hazard testing of materials and reactions.
- Hazard analysis studies to review the process chemistry and engineering practices, including thermodynamics and kinetics.
- Examination of preventive maintenance and mechanical integrity of the process equipment and utilities.
- Worker training.

- Development of operating instructions and emergency response procedures.

7.1.2.2 Oxygen-Deficient Atmosphere

Guidelines recommend the following actions on this area:

- Design and placement of nitrogen venting systems according to industry standards.
- Installation of automatic Emergency Shutdown System that can detect and warn of the uncontrolled release of nitrogen (including presence of oxygen-deficient atmosphere in working areas), initiate forced ventilation and minimize the duration of release.
- Implementation of confined space entry procedures, as described in the General EHS Guidelines with consideration of facility-specific hazards.

7.1.2.3 Chemical Hazards

Release of various chemicals from the main process upset, effluent or during handling inside the process plant or in offsite storage/handling areas are to be addressed in the operation system of the refineries. As such, these are being taken care of by process licensors, chemical vendors who have conformed to meet MSDS (material safety data sheet) of the chemicals, as stipulated in statutory obligations and as being followed by the refineries. Various chemicals used in the refineries are chlorine, methanol, MTBE, TAME, corrosion inhibitor, pour point depressant, anti fouling agent, octane improver, pyrolysis gasoline, lubricity improver and so forth. Chemical hazards should be managed based on the results of a job safety analysis and industrial hygiene survey and according to occupational and safety guidance provided in the General EHS Guidelines provided by IFC. Protection measures include workers trainings, work permit systems, use of personal protective equipment (PPE) and toxic gas detection systems with alarms.

In current times, instead of directly using chlorine cylinders, refineries use chlorine generators for on line dosing thus enhancing safety. As in that example, for every chemical handling, refineries have developed a standard operating procedure (SOP) as per plant operation excellence, discussed in Chapter 3.7. Generation of VOC, PAH, etc., coming out of processes are also to be addressed. Use of proper PPES are vital in chemical handling.

7.1.2.4 Fire and Explosions

This issue is of critical importance given the main processes of refineries are concerned with using flammable hydrocarbons. Various fire and explosion models have been discussed in Chapter 3.2. However, as mentioned by IFC, it is worth mentioning that methane, hydrogen, carbon monoxide, and hydrogen sulphide may ignite even in the absence of ignition source if their temperatures are higher than their auto ignition temperatures of 580, 500, 609 and 260 degree centigrade respectively.

Guidelines recommended the following measures to prevent and control fire and explosion risks from process operations:

- Designing, constructing and operating petroleum refineries according to international standards for prevention and control of fire and explosion hazards, including provision for segregation of process, storage, utility and safe areas. Safety distances can be derived from specific safety analyses for the facility, and through application of internationally recognized fire standards.
- Providing early release detection, such as pressure monitoring of gas and liquid conveyance systems, in addition to smoke and heat detection for fires.
- Evaluation of potential for vapor accumulation in storage tanks and implementation of prevention and control techniques (e.g., nitrogen blanketing for sulfuric acid and bitumen storage).
- Avoiding potential sources of ignition (e.g., by configuring the layout of piping to avoid spills over high temperature piping, equipment and/or rotary machines).
- Providing passive fire protection measures within the modeled fire zones that are capable of withstanding fire temperature for a time sufficient to allow the operator to implement the appropriate fire mitigation strategy.
- Limiting the areas that may be potentially affected by the accidental releases through:
- Defining fire zones and equipping them with a drainage system to collect and convey accidental releases of flammable liquids to a safe containment area, including secondary containment of storage tanks.
- Installing fire/blast partition walls in areas where appropriate separation distances cannot be achieved.
- Designing the oily sewage system to avoid propagation of fire.

Guidelines also provide additional recommendation on management of fire and explosion hazards relating to crude oil storage are addressed in the EHS Guidelines for crude oils and petroleum product terminals.

7.1.3 Community Health and Safety

OSHA-18001 describes community health and safety. As discussed in various sub sections of chapter 3, the subject has been discussed like the

Emergency Preparedness program and Disaster management plan (DMP) in which refineries should focus mitigation measures to be taken for community health and safety in addition to actions to be taken in normal operation as described in Chapter 3.7.

Guidelines recommend actions for the management of these issues as presented below and in the General EHS Guidelines.

7.1.3.1 Major Hazards

- Facility wise risk analysis, including a detailed consequence analysis for events with a likelihood above 10^{-6}/year (e.g., HAZOP, HAZID and QRA);
- Employee training on operational hazards;
- Procedures for management of change in operations, process hazard analysis, maintenance of mechanical integrity, pre-start review, hot work permits, and other essential aspects of process safety included in the General EHS Guidelines;
- Safe transportation management system as noted in the General EHS Guidelines if the project includes a transportation component for raw or processed materials;
- Procedure for handling and storage of hazardous materials;
- Emergency planning, which should include, at a minimum, the preparation and implementation of an Emergency Management Plan, prepared with the participation of local authorities and potentially affected communities.

IFC has not provided any benchmark of performance on community health and safety unlike environment and occupational health and safety, discussed above.

While standardizing the refinery performance to a benchmark, IFC keeps provisions to compare the performance on accident and fatality rates; as such they have indicated the following on accident and fatality rates:

Projects should try to reduce the number of accidents among project workers (whether directly employed or subcontracted) to a rate of zero, especially accidents that could result in lost work time, different levels of disability, or even fatalities. Fatality rates may be benchmarked against the performance of facilities in this sector in developed countries through consultation with published sources (e.g., US Bureau of labour statistics and UK Health and Safety Executive).

8

Safety Tips on Operations

Gaining knowledge and experience of incidents and safety practices in unit operations and sharing with stakeholders are important leading indicators in safety performance. Some of the tips areas wise are given below:

8.1 Furnace

a. In furnaces where only gas burners are used, the main burners are always connected with pilot burners. The pilot burner should be first lit up before the main burner to avoid an explosion.
b. In oil burners, maintain a differential pressure between atomising steam and oil. If the steam pressure is lower than oil pressure, then backfire occurs at the bottom of furnace; therefore, no person should stand near the furnace bottom. However, if steam pressure is too high, the burner gets extinguished.
c. To start a furnace operation, the stack damper should be open, and then the FD fan started. The stack damper should be closed after staring ID fan, then the stack damper security is taken online; thus, if ID fan is tripped, the FD fan is also tripped if the stack damper doesn't open. If the above order is not maintained, an explosion will occur. The stack damper security is also connected to the furnace main security like feed failure when fuel cut off also takes place.
d. In a furnace economizer and super heater coils, flow is to be established before the furnace box temperature crosses 250 °C; otherwise, these coils may get damaged due to dry running.
e. In any major leak of the furnace tube after emergency shutdown the furnace temperature should come down below 100 °C before starting coil steam purging inside tubes; otherwise fire will aggravate.
f. To avoid backfire there should be some negative draft inside the furnace box. Please be alert to check the furnace box through a peep hole with a face shield cover to check that the furnace inside is

visible and not smoky. Peep holes should be covered with a glass shield to avoid burning from back fire if the draft becomes positive.

g. If all the tubes' skin temperatures are normal except one or two, calculate theoretical skin temperature or look through the peep hole with a face shield. If yellow or white spots are observed on tubes skin, then remove the burner from that zone. If yellow or white spots are still seen on the tubes skin, then plan for shutdown to replace the tubes.
h. To avoid furnace tube coke up due to low flow, low feed flow security must be inline after establishing feed flow.

8.2 Distillation Column

a. Before commissioning stripping steam, condensate must be drained out to avoid dislodge of column trays/packing.
b. LPD valve and open discharge point should be far away.
c. In distillation column the bottom level of liquid should be below 80% because indications above this may misguide and liquid may reach up to the trays. This may cause dislodge of the trays.
d. If any leak occurs in crude distillation column bottom zone, then for emergency shutdown the column must be evacuated by running all pumps except the bottom pump. This bottom pump can be started to evacuate the column after cooling down the column temperature below 200 °C to avoid firing in the bottom of the column.
e. For vacuum distillation column, the same process is followed; but here, the vacuum is to be broken by introducing stripping steam and finally by opening the inert gas valve at the top of the column to avoid a vacuum by condensation of steam.
f. Please provide one shutdown valve at atmospheric and vacuum column bottom pump suction away from the pump and it should be remote operated to ensure closing of the column bottom in case of fire at the bottom pump. This may happen if mechanical seal leaks or any other leak happens and fire takes place due to low auto ignition temperature of residues. The same concept is to be followed for other columns handling heaving residue at the bottom, for example, Coker distillation column, Visbreaker column, etc.
g. In a vacuum distillation column, if any leak is on the top or middle zone, then column top pressure indication fluctuates and deteriorates slowly if the leak increases; If flash zone pressure does not

fluctuate or deteriorate, leak survey can continue without shutdown, otherwise the unit should immediately be shutdown.

h. If the suction pressure of pump of vacuum distillation column fluctuates, then first stop quench steam on mechanical seal (for hot dirty fluid) and then immediately start the pump and reopen the quench steam,. If it again fluctuates, then the pump must be changed over to standby and next it should be kept isolated; otherwise air may ingress into the column and may cause an explosion.
i. For stripping steam commissioning, first drain condensate from steam control valve upstream, next isolate it, then drain from downstream, then commission.
j. If a leak occurs in thermo well (TW) tapping of connected pipe lines (inlet pipe line coming from furnace or any product pipe line) of vacuum distillation column, then vacuum indication of column must be fluctuating/hunting; the cause of leak may be sheering of TW or leak from joint of TW; TW must be kept at 45 degree with flow direction in pipe, and to avoid a leak from the joint, use proper gasket in the flange.
k. To start operation of vacuum distillation column, all bleeders connected to column must be kept closed after steaming/condensate draining to avoid explosion by air ingress into the column.
l. Keep two level indications at column bottom to cross-check each indication.

8.3 Column/Vessel/Piping in General

a. Isolation valves' wheel of safety valve must be kept in downward direction to avoid detachment of valve spindle while closing during shutdown.
b. Before commissioning pipeline, its spring type support must be unlocked to avoid the damage of pipe lines.
c. Before feeding any hydro carbon into pipe lines, the vent and drain of the pipe lines must be closed by passing steam or water before commissioning.
d. In steam condensate line, if there is hammering sound, drain condensate at low point upstream of stream trap which might be malfunctioning, thus avoiding damage of pipe lines.
e. During any fire from a pipe line leak, immediately shut down unit if leak source is not visible; stop all pumps from electric substation to avoid fire aggravate.

8.4 Vessel

a. When there is pressure increase in a low pressure vessel connected to the bottom of a high pressure vessel, the shutdown valve is to be closed manually because it may be due to malfunction of level indication of high pressure vessel.
b. When the vessel top vapor routed to flare or compressor suction or to any other vessel in the series and liquid entrainment is occurring in these respective downstream equipments, then check whether upstream vessel level increased or temperature increased or its demister pad (if provided) has detached or not.
c. When vessel level is not decreasing though flow rate is normal and liquid is not of a congealing nature, check whether the vessel's bottom vortex breaker has detached and blocked the outlet nozzle or not.
d. Provide two level indications to cross-check each indication.

8.5 Heat Exchanger

a. If there is any gasket leak on exchanger by thermal shock, the exchanger is to be isolated and bypassed or the unit needs to be shut down if it is light hydro carbon; if it is heavy dirty fluid, then a steam hose can be placed on the leak and one can wait till coking up in the zone and stop the leak. This happens during startup sometimes.
b. In a steam heater/condenser if there is a leak from the exchanger, leaking in either side is unsafe. This leak is to be identified before it aggravates.
c. In a water cooling exchanger, cooling water gate valves wheel to be in downward to avoid valve spindle detachment during closing of the valve earlier.
d. For design of heat exchanger with low delta pressure between shell and tube side, provide high delta pressure security to trip the flow at both side simultaneously like unit shutdown where there is no isolation in the system to avoid delta pressure build up. This would avoid damage of the heat exchanger.

8.6 Reactor

a. In Hydrotreater/Cracker/Platformer, maintain H_2/HC ratio and recycle hydrogen gas flow to avoid coking of catalyst; in steam reformer, maintain steam/carbon ratio.
b. Maintain very low specified sulfur level in platformer and steam-reformer to avoid de-activation of catalyst.
c. Quench valve must not be fully open to take care of exotherm to keep it within allowable range.
d. In FCC, delta pressure in sliding valves to be maintained within norms.
e. In FCC, reactor outlet to fractionator inlet valve de-blinding operation during start up should be done following procedure to avoid fire and fatality.
f. All securities check to be foolproof and LPDs to be closed before start up.
g. Rector's additional safety features provided by the licensors should not be overlooked.

8.7 Compressor

a. In centrifugal compressor, where both lube oil and seal oil are common and circulated from the same tank, PCV setting calibration after every maintenance shutdown should be checked with respect to differential pressure between reference gas and seal oil pressure because if PCV setting is wrong, like seal oil pressure kept lower than gas pressure or if the equalization valve of PCV is there and kept open by mistake, then the reference gas may flow reverse to seal oil tank whose safety will pop as well as the tank may explode. The problem would be severe if lube oil and seal oil are common and stored in a single tank.
b. During turnaround, the compressor suction circuit not used for maintenance and inspection should be kept under inert gas blanketing to avoid compressor suction choking and seriously in case of centrifugal and non lubricated reciprocating compressor.
c. Staffing box vent valve to flare provided for reciprocating compressor should be kept with lock open, as closing the same by mistake would cause explosion of the staffing box.

Also, if inert gas is provided to counter seal leakage of gas, that valve on inert gas also should be kept lock open.

d. If the discharge temperature of a reciprocating compressor gets high due to loader valve malfunctioning, the same should be detected by hand touching the valve or using a digital temperature scanner to avoid rupture of the valve seat.
e. A knocking sound from the compressor is unsafe; first attempt should be to check liquid level in first stage suction and all intermediate stage knockout vessel(s) to drain the same. Defective loader valve also may cause the sound.

8.8 Pump

a. Centrifugal pump is to be started with suction valve open and discharge valve in close condition, while positive displacement pump should be started with both suction and discharge valve open for smooth starting in both cases; for positive displacement pump, if it is started with discharge valve in close position, the pump may rupture with the pressure build up.
b. For high capacity positive displacement pump, if vibration occurs, then after checking all aspects as per mechanical engineers and finding them sound, a shock absorber is to be provided.
c. A centrifugal pump should not be run below minimum flow as the pump may get heated, and after long time it may explode.
d. Seal system of the pump is to be first established before starting of the pump to avoid seal damage and leak.
e. For vacuum pumps, both suction and discharge valves should be kept closed during idling unless required otherwise for warm up; this is to avoid air sucking if seal gets defective.
f. Don't run pump with hammering sound for long.
g. API seal vent valve to flare should be kept lock open.
h. API seal system with inert gas seal using differential pressure between pump seal fluid and inert gas and maintaining level of the seal fluid in a pot should be provided with a high level alarm facility to detect seal leak and stop the pump before seal leak aggravate.
i. Sun dyne pumps operate with high impeller speed of about 29000 rpm and above. Hence, the pump should be given a kickstart and stop first

to ensure proper lubrication of the gear box. After one minute gap, it should be finally started.

8.9 Piping

a. Piping with hot services are generally steam traced along with insulation; thus with a leak starting initially it can't be observed; with smoke coming along with one/two drops of liquid helps one to detect the leak; however, that may be too late, particularly for high pressure and high temperature services which may lead to a major fire.

b. It is seen also that reasonable thickness reduction occurs in elbow zones resulting in leaks. Also in the locations where branch piping are made with a reinforcement pad, leak occurs from the eyeholes of the pad which is a result of poor welding in making stub in the branch joint. The problem is that in stub joint radiography is not feasible and hence, one has to do dye penetration test with each layer of welding; however, this test at each layer of welding is very time consuming because after each phase of welding, one has to cool the joint before doing a dye penetration test; hence, people generally do dye penetration only after the first layer of welding like other conventional butt-joints as carried out in co-axial pipe joints. Hence, the last resort of safety is a hydro test after complete welding when such stub joints are to be specially inspected visually, and where reinforcement pad is required to be welded as discussed above, It should not be done after this hydro test, rather the hydrotest should be done after the reinforcement pad is welded.

c. A leak is also found to occur in sour gas circuit or sour water circuits sometimes. The solution is that one has to take care of proper metallurgy in the design or in the change management if it was not considered in the design. For example, NACE metallurgy must be used, flow rate, temperature, pressure cannot be increased beyond design.

d. Sometimes, capacity revamping done in a unit where many modifications carried out in the units but not in the piping size justifiying that there are cushions in the existing piping. But after some days of operation at revamped capacity, it is observed that thinning in piping occurred, indicating that there might be some lapses in design checks. Hence, the owner should verify this aspect with the designer during revamp project execution.

8.10 Offsite

- *Dispatch facilities*

a. In tank truck/tank wagon loading/unloading, earthling to be ensured before hand to avoid fire by static electricity; follow same for tanks/lines.
b. There must be adequate platforms above the containers of loading trucks and wagon rakes and operator should never cross the platform to reach oval shape of the container as accidents with fatality occurred by slipping from the top of the container.
c. Loading velocity should be kept below 6 m/s to avoid static electricity generation and resulting in fire.
d. The loading/unloading trucks should be fitted with spark arrestor and the same should be checked whether it is working or not.
e. Loading/unloading truck cannot be allowed to kick start its engine before checking the site with explosimeter.

- *Storage tanks, bullets and Horton sphere*

A brief introduction is given below in this regard, as the word, 'storage tank' should not be oversimplified like general storage tanks.

Storage tanks are used for storing petroleum products, process chemicals and also for any liquid storage. There are two types of storage tanks; namely fixed roof and floating roof. In current days a third category has come into being, i.e., a fixed cum floating roof which is a floating roof tank with one more fixed cover over the floating roof deck.

The fixed roof storage tanks remain in atmospheric pressure while floating roof tanks remains marginally above atmospheric pressure due to volatility and vapor pressure of the storing hydrocarbon or chemical. The floating roof deck balances with the atmosphere by floating on the liquids based on level of the liquid; in doing so, the periphery of the floating roof deck with the internal shell wall of the tank is sealed through specially designed compressible material to avoid escaping of the volatile hydrocarbon or chemical to atmosphere. Hydrocarbons, chemicals are classified with respect to their flash points; namely class A, B, C and other category where class A and B categories are stored in floating roof tanks and class C or other categories are stored in fixed roof tanks.

The storage tank like appearance with fixed roof without any nozzle connected with atmosphere and fitted with pressure relief valve(s) falls under pressurized vessel category used for very volatile liquid. However, for liquids which are vapor at atmospheric pressure are stored as pressured liquid in bullet-shape pressure vessels or sphere-shaped pressure vessels (Horton sphere) for large capacities to reduce storing cost.

- *Safety tips for storage tank, bullets and Horton spheres*

a. Tank storage temperature should not be close to 100 °C to avoid boiling, i.e., BLEVE resulting in collapse of tank and fire in the vicinity.
b. In fixed roof tank, storage temperature must be kept below the flash point of the liquid; however, it is safer to follow the rules in a floating roof tank also to avoid it catching fire through chance leakage of vapor through the floating roof deck seal.
c. In a fixed roof tank, in the top vent, where flame arrestor is placed inside, the open end should be covered with a net to avoid bird nest formation.
d. In a floating roof tank, unless it is extremely urgent, de-floating of the floating head by emptying out the liquid to a minimum level should be avoided as in this habit, sometimes a floating roof is found inclined while floating after filling the tank, resulting in liquid escape above the floating roof.
e. Two level indications should be provided with tanks to avoid doubt in one reading.
f. Tank should not be filled above safe filling height.
g. Jet mixer should not be used at very low level to avoid liquid shot into vapor space causing static electricity.
h. Tank filling velocity, particularly in a fixed roof tank should be kept below 6 m/s to avoid static electricity generation.
i. Before running jet mixer or side entry mixer and before starting tank circulation, water, if any, should be drained off to avoid water coming above oil layer and also static electricity generation in case of using jet mixer.
j. The roof drain of the floating roof tank moves through swivel joint and there is a valve at its open outlet outside the tank. The valve always should be kept open; however, sometimes due to swivel joint leak, hydrocarbon liquid drains out; to avoid such loss, operators keep the valve closed which is risky as the floating head may collapse on water accumulation during a heavy shower. In this context, it may be mentioned that there are two drain valves at the tank farm to evacuate its accumulation; namely one outlet connected to effluent treatment plant (ETP) and the other is connected to storm water drain system; if the tank roof drain valve is to be closed due to a higher leak in the swivel joint, then the tank farm drain can be kept open to the storm water side and whenever it starts raining, operator has to go to open the roof drain valve

immediately which should be a practice. In a tank farm where there is no or minor roof drain pipe leak, its valve should be kept in open position and tank farm drain valve should be kept open to ETP side but when it starts raining, valve to storm water side should be made open to avoid overload in ETP.

k. Earth pit connection to each tank should be monitored from time to time to take care of situations like lightening.
l. While floating roof tanks are fitted with gas detector, every tank farm in present times is provided with a gas detector; the operability of the same should be checked from time to time.
m. In LPG Horton sphere, bullets, water draining is done before its weathering quality test. In such draining facility the open end of the drain pipe should be far away from the valve so that when after water draining is over and LPG vapor starts escaping, the operator can quickly close the drain valve.
n. Horton spheres and bullets should not be filled above a safe filling height.
o. Hand torch for lighting used in tank farm in night should be flame proof; and that's why it is called safety torch.

8.11 Quality Assurance System

a. Many piping insulated with or without heat tracing, water cooling heat exchanger with shell insulation are found to be affected with under insulation corrosion over a period of time. It happens very slowly due to damage of the insulation. Hence, insulation of the cooler should be removed, well painted and provided with steel nets to have a human shield to avoid injury. For piping, the same can't be allowed. Hence, there should be a program of random checking of the pipes by removing a small section of the insulation and accordingly making the pipe replacement plan.
b. Sometimes an accident by leakage from a pipe occurs during operation; it is found thinning at the elbow zones, which particularly happens when revamping of the unit is done to operate the unit at higher feed rate when some small areas like elbows, etc., are missed out. Hence, before revamping, the owner should make a checklist of all small items, particularly with respect to quality assurance of the project so that design/engineering agency can be highlighted on the same. Also, there should be adequate length gap between an

instrument in the pipeline and change in flow direction, i.e., bending of pipeline. This gap should be as per standard engineering practice to avoid failure in the bend.

c. Sometimes, hydrocarbon fumes are observed coming out from thermo well pad or from thermo well joint in the piping. It is found that while tag welding thermo well pad into the main pipe, the welder might have pierced the main pipe and quality assurance on the piping concluded before joining this pad. The owner should take responsibility of the lines after hydro test is over and permission should be taken from the owner to carry out any job on the hydro tested line.
d. For leaks through thermo well joint, it is found that either thermo well has sheared or thermo well fit to nozzle was not proper. Again this is an ownership issue; the owner's operating crew should take ownership of fitting all small bore items as mentioned above. Similar problems occur in fitting of orifice taping.
e. Sometimes, wrong gaskets are fitted across the flanges causing leaks. Once, asbestos gasket fitted in flanges having operating pressure of 10 bar and temperature of about 250 °C in one unit start up when, however, the temperature was as low as about 70 °C before the furnace lit up; suddenly, the gasket gave away, and a person moving nearby was hit by the leaky fluid pressure and fell down on the ground; luckily as the temperature was low, he did not receive a major burn.
f. When golden joint (last welding joint without hydro test) is allowed in piping due to its long length and thus difficult to take huge load during hydro test, as per norms, radiography should be done for root welding and for each subsequent layers of welding; owner should oversee that this is religiously followed.

8.12 Miscellaneous Safety Tips

a. H_2S detector should be used to enter inside the area where there is chance of sour gas presence like sour water absorber, amine absorber, compressor using gas which may contain H_2S as impurity, etc.
b. Bolts with shorter length in flange connection are to be detected to avoid flange leak due to weak strength of joints with shorter length bolt.
c. Don't execute online flange leak tightening for liquids of class A and B as well as for vapour to avoid fire by spark during tightening.

Even using a non-sparking tool can't save from fire because there are many zones in the flange to cause friction during tightening.

d. Explosimeter should be used before starting hot work.

e. In open draining from LPD, operator should not leave the place until closing the drain valve.

f. There should not be any open drain facility for sour water or sour liquid; it should be to CBD; if there is a chance of drain line choking, provide flanged lines with isolation so that the line can be dechoked. Similarly, there should not be open venting of sour gas; it should be connected to a flare; if there is chance of vent line choking, provide flanged line with isolation so that line can be dechoked after isolation.

g. Whenever leaving an office room unmanned, switch off the air conditioner (AC), as many a time the capacitor condenser of AC caused fire by short circuit.

h. In any open draining system, the drain valve should be far away from the open mouth of the drain line to avoid a gas cloud when liquid draining is over. Also, the operator should not leave the place keeping the drain valve open until draining is complete.

i. One should not try to de-congeal a drain line by rod poking through drain bleeder as there can be a sudden release of the choke causing gushing out of the holdup liquid which on splashing may cause burn injury, if the liquid is hot. It should be done by steam heating and that drain valve should be far away from the open mouth of the drain.

j. In electric circuit breaker panel maintenance, e.g., MCC (Motor control centre) panel, breaker is taken out after de-energizing the panel; after maintenance while putting back the panel into the cabinet slot, flash fire occurred by short circuit resulting in a fatality of the workman. To avoid this, physically the cabinet slot is to be checked by a second person, remove the dust, check the lever and switch to find out whether there is a chance of short circuit while putting back the panel into the slot; then a final decision should be taken to put back the MCC panel into the slot of the cabinet.

k. For all PCC and MCC rooms in electrical sub-station, to prevent dust, man must not be allowed to enter with shoes; also the rooms' windows and doors should be kept closed.

9

Incident History and Case Studies

From the advent of industrialization until the present day, there have been important advancements in the operation of the oil and gas industries and there have been numerous developments in safety features in response to major accidents, such as the emission of toxic gases, fires and explosions. More advanced technologies with regard to safety features might have enabled us to adventure into more hazardous operations to achieve greater competitive advantage in the market while also achieving a reduction in accidents, but this remains an open question. Earlier, the process was simpler. Hazard mitigation techniques were less and accidents occurred; however, in current times, although there are many risk mitigation techniques in place to avoid risk, the increase in hazardous processes being undertaken by the industries, render these risk mitigation measures often inadequate in preventing accidents. It is observed from a review of the industrial accident history that not a single year has passed without an industrial accident.

9.1 Incident History

Following is a short history of accidents since the early days of industrialization. In the list, there are few reports of incidents in industries outside the United States, and little data for India is available. Note that the accidents for which fatality or injury figures are not available have been omitted from the list.

Industrial explosion

Industry Location	Chemical Used	Year	Fatality, nos.	Injury, nos.
Ohio, USA	LNG	1944	130	-
Ludwigshafen, FRG	Dimethylether	1948	245	3800
Bitberg, FRG	Kerosene	1954	32	16
Louisiana, USA	Isobutane	1967	7	13
Pernis, The Netherlands	Oily slops	1968	2	85
St. Louis, Illinois, USA	Propylene	1972	8	230
Decatur, Illinois, USA	Propane	1974	7	152
Flixborough, UK	Cyclohexane	1974	28	89
Texas, USA	Gas oil	1975	5	-

Beek, The Netherlands	Propylene	1975	14	107
Maryland, USA	LNG	1979	1	1
New Mexico, USA	NG	1983	-	2
San Juanico, Mexico	LPG	1984	1099	-
Vizag, India	LPG	1997	25	-
Panipat, India	Hydrogen	1999	7	-
Utah, USA	Naphtha	2004	-	4
Raleigh, North Carolina, USA	HC	2006	3	1
Texas, USA	Propane	2007	-	3
Texas, USA	HF	2009	-	1
Jaipur, India	HC	2009	12	-
Tinsukia, India	HC	2009	2	-
Vizag, India	LPG	2013	22	-
Louisiana, USA	Isobutane	2016	-	6
LG, Vizag, India	Toxic gas	2020	11	5000
Major fires				
Cleveland, Ohio, USA	Methane	1944	136	77
Port of Texas, USA	Ammonium Nitrate	1947	581	-
Feyzin, France	LPG	1966	18	90
Staten Island, New York, USA	LNG	1973	40	-
Santa Cruz, Mexico	LPG	1985	650	2500
Texas, USA	HC	2005	15	180
Manali, India	HC	2008	2	2
Digboi, India	HC	2009	2	-
Gujarat, India	HC	2009	-	13
Washington, USA	HC	2010	7	-
Vizag, India	HC	2014	4	39
Mathura, India	HC	2020	1	-
Toxic release				
Poza Rica, Mexico	Phosgene	1950	10	-
Wilsum, FRG	Chlorine	1952	7	-
ICMESA, Seveso, Italy	Dioxin/TCDD	1976	o.8 lakh (animal)	
Cartagine, Columbia	Ammonia	1977	30	250
Baltimore, USA	sulfur dioxide	1978	-	100
Chicago, Illinois, USA	hydrogen sulfide	1978	8	29
Bhopal, India	Methyl Isocyonate	1984	2000	20,000

It has also been found that the majority of the accidents occurred due to human error. In incident case studies analysis, as per report of OISD (Oil Industries Safety directorate), India, in 2009, it was reported that about 45% of accidents occurred due to equipment failure and miscreants acts and about 50% of accidents occurred due to human error, with 67% of human error due to non-adherence to SOP (Standard Operating Procedure).

9.2 Accident Case Studies in Petroleum Refineries

As many as 84 case studies of cause analysis have been covered in this chapter. When education is the main objective, effort was not taken to include details such as the name of the refinery and date and time of the accidents. It is seen from the case studies below that in the majority of cases the cause was human error, and in some cases, the accidents occurred due to a lack of a knowledge sharing process among the industries which led to the repetition of similar human errors. The case studies are categorized in terms of nature of accidents as follows:

9.2.1 Case Studies Inside India

a. **Fall from height:**
   It is found from following that about 20% of the accidents were due to fall from height. There are international standards to follow procedures under the category of workplace safety. Also, in India there is an OISD standard-192 to follow procedures for working at height. Even then, the accident frequency is high. This could be due to a need in improvement in the organizations' administrative policies and oversight. Let us now describe the case studies in detail to understand the problems.

   1. **Incident: Fall from height inside APH duct.**
      Description:
      CDU-II during turnaround and inspection, the contractor's welder, along with his helper, was working inside the duct of an air preheater (APH). The helper, while moving with a light bulb holder inside the duct accidentally fell from height about 4.5 meters into a vertical opening leading to the drop out door and succumbed to his head injuries.

      Cause:
      Non-adherence to precautions required during working at a height/confined space. Sometimes it happens that workers, while moving from one place to another at height, unlock the safety belt first and then move to the second location, then again lock the safety belt at the new location. This was the case and the worker fell down in the intermediate location without his safety belt.

      Lesson(s):
      - Job safety analysis (JSA) should be carried out prior to commencement of the job to identify operational hazards

involved and required safety precautions like use of PPEs, safety belt, safety harness, arrestor net at ground level, etc.

2. **Incident: Fall from shuttering of building under construction.**
   Description:
   In the construction site of a new administration building, a contractor employee was walking on the shuttering prepared for the first floor slab of the building under construction. One of the shuttering plates gave away as the worker stepped on it. He fell on the floor below and succumbed to a head injury in the hospital after a few hours.

   Cause:
   Improperly secured shuttering plates due to lack of supervision.

   Lesson(s):
   - Supervision at project site should be ensured.

3. **Incident: Fall from height of a new flare line.**
   Description:
   While removing a spade between two isolation valves of a new flare line, a contractor employee working on the job at a height of 10 meters fell from the pipe rack and subsequently succumbed to injuries.

   Cause:
   Work permit conditions were violated by the employee as he removed his safety belt and fresh air mask just before the incident. Deficiency in scaffolding and working platform provided for working at height.

   Lesson(s):
   - Work permit conditions must be adhered to.
   - Construction of scaffolding should be in line with standard engineering practices and with proper space for safe movement.
   - Competence/skill of workmen should be ensured for safe execution of the activity.

   Note: Use of a safety harness should also be followed to enhance safety for working at height.

4. **Incident: Fall from height.**
   Description:
   While carrying out scaffolding work in FCCU for fireproofing at a height of 12 meters, the scaffolding pipe slipped. The worker, who was wearing a safety belt, fell down along with the pipe. He suffered multiple fractures and succumbed to injuries.

Cause:

Scaffolding was not 'fit to use'. It should have been checked before use which was an example of a lack of supervision.

Lesson(s):

- Scaffolding used should be 'fit to use' and to be tagged after inspection.
- While a safety belt should be used, a safety net also should be provided.
- Supervision should be ensured.

5. **Incident: Fall from height.**

   Description:

   Fall from height while erecting scaffolding at a cooling tower in a Lube refinery resulting in fatality of contractor employee.

   Cause:

   PPE was not used; lack of supervision.

   Lesson(s):

   - Proper procedures and practices should be followed for scaffolding erection including the use PPEs and a safety net.
   - Lack of supervision.

6. **Incident: Fall from height by slippage on staircase.**

   Description:

   A contract worker, while climbing a staircase for conducting a radiography job on a technical structure, slipped and fell down. Subsequently, he succumbed to injuries.

   Cause:

   He might have been unmindful in climbing the staircase.

   Lesson(s):

   - Handrails should be used while moving on a staircase.
   - Safety briefs should be given while attending duty so that safety awareness, including not to attend job when in poor physical health or in poor mental condition.

7. **Incident: Fall from height.**

   Description:

   In Visbreaker unit, one technician fell down from a monkey ladder and succumbed to injuries.

Cause:

He might have been unmindful while climbing/coming down the ladder, hand gloves were not up to the mark or monkey ladder rods were oily—which might have been caused by the previous person with shoes patched with oil.

Lesson(s):

- A complaint should be lodged against the monkey ladder conditions instead of using it without delay.
- A safety awareness briefing should be provided before joining daily duty, and workers should be reminded to refrain from duty when in poor health or adverse mental state.

8. **Incident: Fall from height.**

Description:

In PX-PTA project of a refinery, one fatality occurred inside CCRU furnace due to falling from height (19 meters) during construction activities.

Cause:

Worker entered an unauthorized project area.

Lesson(s):

- A daily safety awareness brief should be given before the start of work and workers reminded to refrain from unauthorized entry to a work area.
- Supervision should be improved.

9. **Incident: Fall from roof of LPG filling shed.**

Description:

Contract workman fell from the roof when old asbestos sheet gave away and succumbed to injuries.

Cause:

Use of safety features like PPE, hand gloves, safety belt/ safety harness and safety net were not ensured before issuing the job permit.

Lesson(s):

- Adequate supervision and safety awareness needs to be ensured regarding work at height.
- Job safety analysis is to be done and actions taken before issuing a job permit.

10. **Incident: Fall from height during lifting of a pipe.**
Description:
At one project site, a contract worker (rigger) fell from a 3 meters height due to an imbalance caused by a wobbly pipe of 2.5 meters length while it was being lifted up. He succumbed to injuries.

Cause:
Safety gadgets not used while working at height.

Lesson(s):
- Adequate supervision and awareness to be ensured regarding working at height.
- Safety briefs to be provided before job start.
- Use of safety gadgets like safety belt, safety net as per requirement must be ensured.

11. **Incident: Fall from height.**
Description:
The contract worker slipped from 4.2-meter height of scaffolding at Co-Generation plant and succumbed to injuries.

Cause:
Guide rope/lifeline was not used for anchoring safety belt.

Lesson(s):
- Adequate supervision and awareness to be ensured for working at height.
- Safety brief to be given to use of safety gadgets like safety belt, safety net as per requirement must be ensured.

12. **Incident: Fall from height during gas cutting in project area.**
Description:
Contract worker fell from 9-meter height while removing a redundant structure by gas cutting and subsequently succumbed to injuries.

Cause:
Guide rope/life line was not used for anchoring the safety belt.

Lesson(s):
- Adequate supervision and awareness to be ensured for working at height.
- Safety brief to be given in the use of safety gadgets, such as safety belts and safety nets, as per requirements.

13. **Incident: Fall from height from godown roof top.**
Description:
Contract worker fell from 8 meter height while laying telephone cable on the roof top of a maintenance godown when an asbestos sheet of rook had broken.

Cause:
Adequate fall protection was not used.

Lesson(s):

- Adequate supervision and awareness to be ensured for working at height.
- Safety brief to be given in the use of safety gadgets such as safety belts and safety nets, as per requirement.

14. **Incident: Fall from partition wall of building.**
Description:
While laying electrical cable on the false ceiling of a new building block, a contract worker fell from a partition wall (approximately 2 meters high) and succumbed to injuries.

Cause:
Construction safety rules were not followed.

Lesson(s):

- A safety brief should be provided before job start.
- Supervision should be strengthened.

15. **Incident: Fall from height; namely from the top of BTPN wagon (higher capacity wagon).**
Description:
At sixth gantry of an OMS (Oil Movement and Storage) area during rake loading, an operator slipped from the BTPN wagon top and succumbed to injuries.

Cause:
In the absence of safety belt provision for the loading operation, he fell due to an imbalance in the curved surface while handling a hose.

Lesson(s):

- Like that provided in gantries in other refineries, provision of hydro rails should be considered so that it can be held by the operator for support during a wagon loading operation.

16. **Incident: Fall into storm water canal.**

Description:

While unloading quarry dust, one of the material supply contractor workers walked over a culvert parapet wall and fell into storm water canal. He was taken to hospital where he died after ten days of treatment.

Cause:

It was a case of lack of perception of a safety hazard.

Lesson(s):

- Activity should be covered under the work permit system and alertness should be inculcated in the workmen while moving in the area.

b. **Road accidents:**

It is found that there has been about 10% road accidents though there is speed limit of 20 kilometres an hour in the inside roads of the installations of the industries. The details are shown as follows:

1. **Incident: Accident by water tanker.**

Description:

A contractor employee was taking a rest after lunch under the shade of a tree away from the road on the north side of a permeate tank of TTP plant. A tanker lorry lost control while taking a left turn. It crossed the roadside storm water channel and hit the tree branches, hitting the victim underneath and finally the concrete wall of the water tank. The victim who suffered severe head injuries was declared dead after arrival at the hospital.

Cause:

Non-adherence of traffic rules inside the refinery.

Lesson(s):

- Traffic guidelines circulated by OISD should be enforced for road safety within battery area/plant premises.
- Speed control within refinery to be enforced.
- Administrative action to be taken against traffic violations and erring transporters and pedestrians.

2. **Incident: Road accident by POL tank truck.**

Description:

A contract employee was working on the ground along the left half of the road towards the west gate 2. It appeared he was walking on the extreme right of a group; i.e., almost in the middle of the road. A lorry coming from the loading bay side was approaching the west gate 2 in the same direction, and the side of the lorry hit the victim, who then fell down near the rear wheels and was injured in the hips as he was pressed by the wheels. When taken to the hospital after first aid was administered, he succumbed to injuries.

Cause:

It was due to lack of alertness in driving and violation of rules by the pedestrians.

Lesson(s):

- Traffic guidelines circulated by OISD should be enforced for road safety within battery area/plant premises.
- Alert driving with speed control to be enforced inside the refinery.
- Administrative action to be taken against traffic violation and erring transporters and pedestrians.

3. **Incident: Road accident by bus inside the refinery.**

Description:

While crossing the road inside the refinery, an employee was hit by the bus. Though the bus driver applied the brake, the employee fell on the road and sustained head injuries. He was shifted to the hospital where, after a few days, he succumbed to injuries.

Cause:

It was due to violation of traffic rules inside the refinery.

Lesson(s):

- Traffic guidelines circulated by OISD should be enforced for road safety within battery area/plant premises.
- Speed control within refinery to be enforced.
- Administrative action to be taken against traffic violation and erring transporters and pedestrians.

4. **Incident: Worker run over by LPG tank truck.**

Description:

Near the main gate of LPG dispatch unit, a contract security guard was run over by an LPG loaded tanker which was moving towards Weigh Bridge.

Cause:

It was due to brake failure of the vehicle and there was a downward slope at the gate.

Lesson(s):

- Better maintenance of vehicles with periodic inspection of brakes must be ensured.
- Road safety rules must be adhered to.
- Drivers should not take the driving seat until the gate is completely open. Traffic guidelines circulated by OISD should be enforced for road safety within battery area/plant premises.

5. **Incident: Road accident in refinery.**

Description:

While taking a U-turn, one truck hit a laborer, who succumbed to injuries in hospital.

Cause:

Non-adherence to traffic safety rules.

Lesson(s):

- Road safety rules must be followed.
- Training, awareness and adherence to traffic safety rules to be followed.

6. **Incident: Careless driving inside the refinery.**

Description:

Jeep fell into a tank dyke open space while driving as the jeep hit the parapet of the road during movement. This resulted in more than 500 accident-time man hours lost.

Cause:

It was a case of careless driving.

Lesson(s):

- Drivers should be more attentive and careful while driving.
- OISD guidelines on driving should be followed.

7. **Incident: Worker fatality due to being run over by hydra crane.**

Description:

While shifting the structural member of a sluice gate to the pond using a hydra crane during its fabrication by contract workers at plant area, a rigger fell down and got struck by the front wheel of a hydra crane. He was shifted to the hospital where he was declared dead.

Cause:
Lack of supervision.

Lesson(s):

- Better supervision and on-the-job training is required.
- On-the-job safety procedures for crane driving in co-ordination with riggers is a very important procedure as many times similar accidents have taken place.

8. **Incident: Fatal accident when truck ran over the victim.**
Description:
The driver of an asphalt truck was run over by the truck while he was trying to start the truck while lying below it. He was trying to fix a short-circuiting wire problem that was not starting with a normal ignition key.

Cause:
Testing was done without putting a proper lock mechanism on tire movement; moreover it was the job of a technician, not the driver.

Lesson(s):

- Ensure all systems in truck in good condition before entering the refinery.
- The supervisor should take the lead in preventing such work being done by the driver.

c. **Failure of maintenance machineries:**
About 8% of accidents are due to maintenance machinery failure. Case studies are as follows:

1. **Incident: Toppling down of a crane.**
Description:
During erection, the crane got toppled; the suspended column swayed and fell on the ground. The contract laborer succumbed to injuries as he got entangled under the falling vessel.

Cause:
Mishandling of machine and equipment, lack of planning and supervision.

Lesson(s):

- Proper planning should be done to assess the load and avoid overloading of crane.
- Strict supervision to be ensured.
- on-the-job training to be strengthened.

2. **Incident: Crane failure causing damage to equipment in Aromatic recovery Unit.**
Description:
DEMAG crane failed while lifting a portion of a stripper column for replacing its bottom portion during shutdown. It caused damage to the crane boom, nearby vessels, platform structural and pipelines. There was no injury but damage caused huge financial loss.

Cause:
There was deficiency in crane management.

Lesson(s):

- Crane management should be strengthened by developing skills and awareness among the crew for efficient and safe crane operation and maintenance.
- Any such job must follow strict procedural compliance in checklist form with respect to crane capacity, equipment validity, crane operator's competence, site conditions and execution planning.

3. **Incident: Hit by a falling pipe due to crane slings failure.**
Description:
During erection of radiant section outlet header (28″ diameter × 12.2 meters long) of CCRU (Continuous Catalyst Regenerating Reformer Unit) heater, slings of 75 MT crane got sheared while maneuvering the header pipe to its position. The falling rope hit two workers. One succumbed to injuries.

Cause:
Lapses included visual checking not done on slings condition, wrong choice of web sling for the job; packaging not used to protect the web sling from rough contours with inadequate room for maneuvering and inadequate supervision.

Lesson(s):

- Use slings of proper size and capacity.
- Inspect slings before use and place 'Fit to use' tag on it if ok.

- Involve only authorized supervisors.
- Ensure training on usage and inspection of slings.

4. **Incident: Hit by a rod during bundle lifting by a winch.**
   Description:
   At one project site of a refinery, for the construction of 88 m RCC stack for new boilers, one contract workman was operating a winch to lift a bundle of reinforcing iron rods. As the bundle of the rods got lifted, one of the rods fell down and pierced the helmet of workman causing fatal injury.

   Cause:
   Winch operator was not adequately trained for the specialized job. Job safety requirements were not examined before start of the job and non-standard tools/tackles were used. Site supervisor was ineffective.

   Lesson(s):
   - Only skilled and authorized operators should be allowed to operate lifting equipment like crane, winch, etc.
   - Method statement duly approved by PMC and the owner should cover detailed procedures for safe execution of the job.
   - Lifting system should be duly inspected and certified.

5. **Incident: Exchanger fell down from height during erection.**
   Description:
   In FCC during erection, De-butanizer pre-heat exchanger fell down from 3 meter height due to snapping of the two slings. The proprietor of the executing company got trapped and succumbed to injuries.

   Cause:
   Wire rope slings got snapped as these were not having adequate design capacity of handling the load of the equipment.

   Lesson(s):
   - Ensure adequate capacity of hoisting equipment and matching certificate.
   - No one should go underneath the hanging load.
   - Comprehensive erection plan to be made.
   - Job knowledge of erection equipment to be ensured for execution group.

6. **Incident: Tripod malfunctioned and winch operator hit by a sheared piece of an object.**

   Description:

   At a SPM project site during shifting of the tripod of the piling rig, the connecting pin at the top sheared and one of the tripod legs fell on the winch operator of the piling rig who succumbed to injuries.

   Cause:

   Periodic inspection and repairs of tripod were not done.

   Lesson(s):

   - Standard norms for periodic inspection and repairs must be ensured and records should be maintained.

d. **Explosion and fire:**

Explosion and fires are the most anticipated hazards due to gaps in engineering and operation mistakes that happen in the refineries. It is found that about 35% of accidents occurred by explosion and fire. This indicates the need for further strengthening of safety in the engineering and operation process and this should be continuous. Sometimes explosions and fire result in huge property loss and operation days lost in addition to human fatalities; hence, there is a necessity of structured hazard analysis on the main processes. There is a need to always take the driver's seat when it comes to safety, as discussed throughout this book. To enhance knowledge on the subject, case studies on explosion and fire are narrated as follows:

1. **Incident: Fire in catalytic reforming unit (CRU):**

   Description:

   The vertical submerged feed pump mechanical seal started leaking. The engineer from control room observed some fumes in the outside; walked out of the control room, along with the field operator, and approached the pump. However, by the time they reached the pump area, naphtha vapors from the leaky pump connected with the nearby fired heater and an explosion resulted. The unit was brought to emergency shutdown but both the engineer and operator were engulfed in the fire and succumbed to burns after taken to the hospital for treatment.

   Cause:

   As naphtha is a very light and volatile liquid and creates vapor fumes as soon as it come into open air and the fired heater is in operation in a running unit, the operating persons should have first done an emergency shutdown from the control room by cutting off the fire in the heater and next cutting off feed through a remote switch instead of reaching the leaky pump area.

Lesson(s):

- The incident took place in the early days, when there was a single seal in the pump; whereas in present times for all class A and B hydrocarbons double seals are provided and with an advanced API seal mechanism; i.e., if seal leaks it goes either to flare or to a seal pot, which on high level alarm trips the pump and feed pump trip meaning stoppage of feed flow to heater would cause heater trip in turn by interlock logic diagram. However, with the present technology currently in vogue, operating crews should enter the field after ensuring that the fired heater has been positively cut off; i.e., if pilot burners are on, those should be cut off first from the control room before entering the field.
- A number of gas detectors should be placed in the ground level at different locations as per hazard study.

2. **Incident: Fire in furfural extraction unit.**
Description:

This unit is meant to allow physical separation/extract aromatic from a vacuum distillate to improve its lubricating property when the ultimate purpose is to produce lubricating oil. In this process, furfural is used as a solvent for separation, which is recovered through distillation for reuse. As both the feed and solvent are of high boiling range, the safety valves of the unit are not connected to the refinery main flare and instead all safety outlets are connected to the blow down vessel which is vertical about 20 meters high but with top end open to the atmosphere; if hot vapor enters the vessel, cooling water flow would start in auto mode to condense the furfural vapor, which then flows to closed blow down vessel.

In such a unit, safety pop occurred but then there came a huge uncondensed vapor from the top of blow down vessel. Operating crew reached the site to check whether blow down vessel cooling flow started or not. By the time they understood that the problem was not in cooling flow but rather due to the vapor, which is very light, not condensing properly, the vapor got connected to the nearby fired heater with a resultant explosion. The operating crew while running away from the site got injured but recovered subsequently.

Cause:

There was a common storage tank in the offsite meant for storing both solvents like furfural and MEK to take care of tank M&I activities. In this process, there was some human error in connecting the right tank to supply furfural to this unit, which resulted in pumping of MEK which is very light solvent with a boiling point of about 80°C only. Naturally, safety valves in the extract distillation/recovery column popped and could not ensure complete condensation in blow down vessel resulting in a big fire.

Lesson(s):

- Dedicated solvent tank should be used for each category of solvent which, however, was implemented subsequently.
- Cooling water flow control to the blow down vessel should be operable from the control room.
- HAZOP study should be done before any change in operation philosophy.

3. **Incident: Fire in propane deasphalting unit.**

Description:

This unit was meant for extracting asphaltene out of the vacuum residue to recover lube distillate from vacuum residue. This unit is also used to recover distillate (if not having lube potential or if use for lube is uneconomical) to process in fluid catalytic cracking unit (FCCU). Pressurized liquid propane is used as a solvent to extract oil from vacuum residue.

A huge fire was observed in the unit which was caused by a mechanical seal leak of the propane pump which got connected to the fired heater of the unit. The unit was brought to emergency shutdown by cutting the feed from the control room and stopping the offsite feed pump and cutting off electric power of all motor drives from the electrical substation, as the pumps and compressor were not approachable and there was no stop switch in the control room. There was a propane vessel at the first floor above the pump in fire. To keep the vessel cool, water sprinklers from the control room were started. A Firefighting crew was called to fight the fire; no human injury occurred, since the crew followed safety methods to tackle the emergency but damage to the asset had occurred.

Cause:

There was no gas detector in the pump house and the prototype mechanical seal, i.e., secondary seal outlet was open to the atmosphere instead of API-52 seal; moreover, fire occurred in the night shift due to slackness in monitoring.

Lesson(s):

- HAZOP study should be done and adequate interlock shutdown facilities should be provided for safe shutdown, as well as for quick isolation and cut off the fuel source to fire.
- Gas detectors should be provided at different locations to pre-alert before fire takes place.
- Standard API seal system should be fitted to the propane pump to avoid mechanical seal failure.

4. **Incident: Fire in CDU (Crude distillation unit).**

Description:

Unit was started after M&I shutdown. When the column bottom reached to desired temperature, a sudden fire was observed in bottom pump area; before the situation understood, it became a major fire spreading upward to a very high level. Unit was brought to emergency shutdown and the column bottom emptying operation started with column stripping steam flow continuing. Some contract labors working at column top for fixing the insulation material became very panicky as fire reach near to that height. In panic, they jumped from that height of about 30 meters and naturally succumbed to freefall.

Cause:

One level indication was not working and instrument maintenance was given the job request to rectify the indication. The instrument taping drain bottom was not capped after completion of the job. The bottom valve of the level troll was passing and liquid naphtha was falling on the ground where the hot column bottom pump was located. The situation created the vapor cloud for explosion and fire.

Lesson(s):

- Open draining from instrument connected to volatile liquid should not be allowed; instead the drain should be connected to the flare; for heavy fluid, the drain should be connected to closed blow down vessel (CBD).
- Unit startup and shutdown processes are unsteady processes; hence, no contract or maintenance service job should be allowed during these unsteady processes; in extreme cases, it should be allowed by a higher authority after discussing the hazard potential.

- The auto start facility in the control room should be provided to start the standby pump from remote and should always be kept in lined up condition in the field. This is to reduce the impact of fire.

5. **Incident: Fire in Solvent De-waxing Unit (SDU):**
   Description:
   In SDU, there are numbers of rotary vacuum drum filters used in parallel for operation; many time, one or two remain out of operation. One filter was lying idle for about a week without operation. One day, there was an explosion in the filter causing bulging of the filter hood (top half of the filter), resulting in production loss due to non availability of the filter for about three months. There was no human injury. It was a case of asset loss.

   Cause:
   The filter vat (bottom half) and hood is made of carbon steel; during operation, the filter remains at low temperature like about (−)10°C. and there remains inert gas blanketing inside the hood. When filter remained out of operation for about a week and with the inert gas blanketing valve kept closed to avoid loss of inert gas through leakage, the warm filter came in contact with air during the period and pyrophoric iron formed during such idling came in contact with air causing a fire.

   Lesson(s):
   - Inert gas blanketing cannot be stopped during idling; instead any deficiency in an inert gas circuit should be identified and rectified to hold inert gas system pressure.

6. **Incident: Fire in Catalytic De-waxing Unit.**
   Description:
   Unit was started after a short shutdown. Hydrogen gas circulation was started as per procedure; heater light up was done as per procedure. While firing was being increased to increase heater coil outlet temperature, the temperature was not rising; all systems in the field were checked, such as the fuel gas valve opening, fuel gas pressure, hydrogen gas flow, its pressure, hydrogen compressor operation, and line up and no abnormality was found. When operating crew started suspecting that the amount of fuel being consumed for only hydrogen gas in circulation appeared to be more unlikely, i.e., very high, suddenly, there was a sound of explosion from the field. The unit was brought to emergency shutdown.

When inspected, it was observed that there was an explosion inside the heater box, and the heater box coils appeared to have become damaged.

The unit was restarted after a major repair of heat coil and was normalized. It was a case of asset loss.

Cause:

There was a heater bypass line for the hydrocarbon feed used until the heater temperature was raised to the desired level. A hydrogen line was connected to the feed line, which was open. But there was a separate hydrogen line to ensure flow to the heater during startup gas circulation which was closed. As the bypass line was open, there was no problem in hydrogen flow but bypassing the heater an unknowingly dedicated hydrogen valve to heater was closed. So, the maloperation could not be understood from the control room and the heater coil ruptured due to starvation of flow through coil while coils were under heat up. The heater skin temperature indication was only two in number, which was also doubted.

Lesson(s):

- HAZOP study in the design was not done properly to provide the right connect to overcome a human mistake in the lineup.
- Adequate numbers of skin thermocouples should be provided to get more indications.

7. **Incident: Fire in DCU:**

Description:

During emergency power and utility failure, backfire occurred inside the furnace box, resulting in blocking of furnace tubes by coking up of the liquid inside the tubes and thus caused production loss of about three weeks for dechoking of furnace tubes and repair. No human injury took place but there was about 15 days operating days lost to revive the furnace fit for operation by recovery maintenance cost.

Cause:

The medium pressure steam header was connected to both the heater coil purging line (though through a check valve to avoid reverse flow) and heater oil burners' atomizing steam line. Due to utility failure, steam pressure came down and also due to the valve check not working, reverse flow of feed (remaining pressurized inside the tubes for a few seconds after the feed flow stopped) occurred; thus the oil entering into the steam header entered the atomizing steam line and caused entry of burner oil inside the furnace box without steam, resulting in an uncontrolled fire inside furnace box causing coking of furnace tubes.

Lesson(s):

- Check valve in furnace tubes purging line located closed to furnace feed inlet line to avoid choking the line up to check valve downstream. So, a remote operated on-off valve should be placed very close to the check valve at its upstream as it may not be approachable during emergency. This on-off valve should normally remain closed. While opening, it should be checked that the furnace tube pressure has subsided to a very low level as compared to steam header pressure.

8. **Incident: Fire in CDU/VDU:**
Description:
At around 5.30 AM before the shift change was over, a filed operator heard a huge sound and observed in the next moment a huge fire originating from the main crude oil booster feed pump. The unit was brought to emergency shutdown but before all concerned came into action along with the fire and safety crew, the fire engulfed the pump house area with flames crossing the overhead pipe rack. All necessary isolations were done at battery limit and all pumps were switched off from the electrical substation. But as there were a number of pipelines in the pipe rack which got ruptured through overheating by fire, the volume of inflammable liquids was high enough to prevent extinguishing the fire even after battery limit isolation. Due to fire for about two hours, many overhead pipes were damaged, resulting in a long recovery period. No human injury occurred.

Cause:
Both the crude booster pumps were running at unit feed capacity of 90%; also, in the shift feed was reduced to 80% but one pump was not stopped and both the pumps were running. As per revamped design capacity, one pump can take care of 75% load max. with respect to motor amperage; hence, at more than 75% capacity the second pump should be run. Hence, while running both the feed pumps, the unit feed rate should have been increased to 95% to 100% to prevent any feed pump from running below turndown ratio; otherwise, at turndown capacity, the pump gets heated which occurred in this case, i.e., pump casing got sheared though the bearing was not jammed and the mechanical seal was found to be sound. To confirm the reason, after startup of the unit from recovery shutdown, the feed pumps alternately were taken to the workshop after a 24 hour- run of two pumps simultaneously at 80%

capacity ,when it was observed that there was a thin cut mark generated inside the pump casing circumferentially. Before the revamp, only one pump was running at 95% to 100% capacity and thus there was no problem. But after the revamp, as a single pump running caused higher motor amperage, workers ran both the pumps without knowing the possible consequence; hence, it was a mistake in revamp design by the designer.

Lesson(s):

- Until the pumps were redesigned to make them multistage instead of existing in a single stage, operation practice was changed by operating the unit below 60–70% when one pump was kept in line or operating the unit at 95% to 100% when both the pumps were taken in line and no operation was done at capacity between 70% to 95%.

9. **Incident: Major fire in Hydro-cracker unit.**
Description:
At about 6.00 PM, one field operator heard a huge sound and immediately observed a long jet of white fog coming out of the overhead hydrogen mixed reactor effluent Fin-Fan cooler area followed by an explosive sound and fire. Unit was brought to emergency shutdown immediately and all necessary isolation was done to contain the fire. Fire was brought under control after half an hour and extinguished totally in the next half an hour. There were four fin-fan coolers in the set and all four were damaged, while one cooler's channel box got sheared, causing leakage of hydrogen and hydrocarbon. The fire also affected structure in the technology platform; however, it did not affect the nearby other equipment. The set of fin-fan coolers of new metallurgy of duplex steel tubes was a replacement of an old set of fin-fan coolers of carbon steel metallurgy as a part of a metallurgy upgrade in order to avoid rust deposition inside the tubes during turn around shutdown for long periods when tubes may come in contact with atmospheric air. However, no human injury occurred.

Procurement of a new set of fin-fan coolers took place over an 8-month period. Fortunately, the old set was not disposed and was lying in the scrap area. It was taken up with a reputed engineering house that supplied the original fin-fan cooler, and a plan was developed to refurbish the old set at the owner's site by replacing it with new carbon steel tubes. This took 2 months and the unit could be restarted with a refurbished carbon steel make fin-fan cooler. Action was taken to purchase a new set of duplex steel make fin-fan coolers after taking corrective actions, as evident below.

Cause:

The decision to upgrade metallurgy was taken following the parent company's similar decision earlier. While lining up an agency to fabricate a new set of coolers, the vendor's credentials were considered as experience for a similar job and to be listed in EIL records. In fact, the vendor was listed in EIL record for manufacturing a fin-fan cooler. When going over the pros and cons, it was found that the vendor had the credential to fabricate a carbon steel make fin-fan cooler under EIL record but not duplex steel make. Hence, while looking for credential in the EIL list, eligibility in making fin-fan coolers was noticed but not with respect to metallurgy. Thus, poor workmanship of the vendor in making a duplex steel fin-fan cooler resulted in catastrophic failure of the equipment.

Lesson(s):

- On a short-term basis, the above action was taken and the unit could be restarted in two months and precautions were taken to select the right vendor for long-term action, like procuring a new set again with duplex steel metallurgy.

Note: The lesson learned from experience at one refinery was not implemented in the other refinery, and thus a similar accident occurred in the second refinery. However, in the second refinery, a large amount of nearby equipment was also affected because the thrust of the explosion was much more severe because the capacity of fin-fan coolers was much higher than in the first refinery.

10. **Incident: Fire in DCU.**

Description:

The fire incident occurred without any fatality or injury. In this case, during operations such as remote switchover of the coking drum operation after completion of the respective cycle of coking, it was observed that the furnace outlet pressure shot up. The unit was brought to emergency shutdown but the pipeline rupture occurred at a heater outlet line causing a big fire.

Cause:

The coker drum switchover instruments did not work, causing pressure shoot up followed by line rupture. The leaky liquid had very low auto ignition temperature (as usual for heavy residual liquid), which in contact with air caused fire.

Author's analysis:

In DCU, heater outlet pressure, i.e., coke drum pressures are normally at a level of 3 to 4 bars; if pressure shoots up, the safety valve provided at the heater outlet takes care. In DCU, interlock shutdown facilities are provided by the designer where heater fuel cut off and feed cut off is provided with higher heater outlet temperature and low feed flow but hardly any cut off is provided with heater outlet pressure high; for pressure safety, it is left to the function of safety valves as provided.

It may be mentioned that DCU feed is generally vacuum residue which is very viscous and congealing and may be dirty as well. So, there may be a chance of clogging of liquid at the safety valve inlet due to inadequacy of heat tracing and/or by the dirt.

Lesson(s):

- While strengthening the instrumentation of the coke drum switchover mechanism, thought also should be given to consider inclusion of a high pressure in trip interlock system.
- Electrical heat tracing for viscous and congealing liquid should be considered at least for a small zone like isolation valve, line flanges where steam heat tracing may not touch the piping properly and may cause discontinuity in heat tracing.
- Provision of rupture disc with isolation valves for dual safety valves with one as standby should be provided in between the isolation valve and safety valve and with a pressure indication in between rupture disc and the safety valve to understand whether the rupture disc has been pierced or not.

11. **Incident: Fire due to naphtha leak from air fin cooler tubes.**
Description:

While atmospheric vacuum unit (AVU) was operating at normal, suddenly a fire broke out in the low pressure air fin cooler in the overhead circuit of atmospheric column in CDU (crude distillation unit). Fire was extinguished in an hour after shutting down the unit for safety considerations. Fire was caused by a leakage of naphtha from tubes of air fin coolers and subsequent ignition.

Cause:

Failure of the fin cooler tubes was attributed to higher fluid velocity inside the tubes, leading to an erosion/corrosion effect of processing slop/wild naphtha ex DHDS (Diesel Hydrodesulphurization) unit, which could be a contributing factor.

Lesson(s):

- Study of required modifications is needed to reduce the fluid velocity in tubes with an acceptable limit, as set in the design.
- Provision is to be given at the earliest for water injection at the inlet of air fin cooler to wash out deposits.
- An alternate processing route for wild naphtha is to be looked into.

12. **Incident: Explosion and fire involving slurry settler in FCCU (Fluid Catalytic Cracking Unit).**

Description:

In FCC unit, a major explosion triggered from slurry settler was followed by a fire, which was extinguished subsequently. Extensive damage was caused to the reactor-fractionator section and main column bottom circuit involving slurry settler of clarified oil. The incident occurred during the process of dechoking of the slurry settler bottom line. There were two fatalities in the incident involving contractor workers who were working near the slurry settler at the time of incident.

Cause:

Instantaneous internal over-pressurization occurred in the settler due to sudden inter-mixing of hot slurry with water, which was either present in the bottom as left over, or entered as line condensate from MP steam line while trying to dechoke the slurry settler bottom circuit.

Lesson(s):

- The operating manual of the unit should be updated to incorporate stepwise details of line up of all circulation loops with simplified sketches. The locations of drain points from where water draining is to be done should be specified for all loops and equipment.
- All operating people should undergo a refresher course on the revised manual, in particular with startup, shutdown and emergency handling procedures. They should thereafter be assessed and retrained if needed.
- In case slurry settler is to be taken in line, the normal course of steaming, gas backing and water draining must be carried out and the settler should be taken in the circulation loop, both from top and bottom side.
- All process modifications and changes in operating procedures should be administrated through a structured

'management of Change (MOC),' as per OISD-178 which includes HAZOP analysis, updates of operating manuals and training of personnel.

- A structured checklist may be institutionalized to monitor and follow up the key and critical steps during startup and shutdown to avoid shortcuts/deviations from laid down procedures.

13. **Incident: Leaks and major fire in FCCU.**

Description:

Due to leakage of flushing oil from pipeline on hot surface of a steam line, major fire took place on run down area of FCCU.

Cause:

It was due to close proximity of steam PRDS (pressure reducing station) valve to unit B/L (battery limit). In addition, leak testing procedures were not followed.

Lesson(s):

- The approved procedures for pressure/leak testing should be followed to ensure circuit tightness during startup. In case of deviations; proper approval should be taken, as per management of change in line with OISD-178.
- The steam PRDS valve located at the grade level and in close proximity to the battery limit valves at FCCU should be guarded suitably to prevent direct impingement of hydrocarbon from any source of accidental leak. The PRDS valve along with associated piping should be insulated to avoid any exposed hot surface.
- Segments of circuits checked in the field and passed in pressure/leak test should be marked in P&ID.
- A matrix should be developed for identified critical segments of piping in a unit/area for quick isolation of leaking segments. Such a matrix will be a handy tool as a ready reference of related valves which need to be closed for emergency isolation (those isolation valves may be located ISBL/OSBL (inside battery limit/outside battery limit)).

14. **Incident: Major fire in offsite pipe racks.**

Description:

Fire occurred while laying new pipeline on overhead pipe rack in an offsite project area in proximity to OWS (Oily Water Sewage). It caused the fatality of two contract workers due to burns.

Cause:

There was a gap in assessing safety hazards and a deficiency in hot work site preparation, i.e., without precautions against release of hydrocarbons from OWS.

Lesson(s):

- Through joint visits site hazards should be identified and precautions put in place before starting a hot job.
- Work supervision should be ensured.

15. **Incident: Explosion and fire in OWS (oily water sewage).**

Description:

In AVU, an explosion occurred in OWS dislodging seven manhole covers and was followed by a fire on the pipe rack. One contract worker died when he was hit by one manhole cover.

Cause:

Non-standard engineering was done for connecting the surface drain to OWS without a liquid seal.

Lesson(s):

- Any oily water drain line from the plant area should be connected to OWS only via a designated inlet pipe, leading to a sealed manhole or through a separate catch basin designed to ensure a liquid seal for preventing reverse flow of downstream gases from OWS. Standard design and engineering practices must be adhered to in order to maintain integrity of the OWS system for overall plant safety as stipulated in OISD-109.
- Management of change (MOC) procedures should also be applied for any change in OWS system, as per OISD-178.
- The numbers of OWS vent points available vis-à-vis requirement should be reviewed in line with standard design and engineering practices.
- In line with stipulations in OISD-105, separate work permits must be issued for each point location in the operational unit. Clearance should be issued at site after ensuring that all conditions are met at the point location.
- Stricter supervision for compliance of safety stipulations must be ensured by the engineering in charge, permitee and contractor supervisor to enable safety on a sustainable basis. This includes safety awareness of contractor workers and supervisors in line with OISD-207.

16. **Incident: Fire at CRU (Catalytic Reformer Unit) reactor inlet utility point.**
Description:
Fire broke out at the third reactor inlet line elbow bend due to a leak at utility point (3″ diameter) through which injection of DMDS (Dimethyl Disulfide), CCL4 (Carbon tetrachloride) and air is done. It took about 90 minutes to extinguish the fire. The unit was brought to shutdown.

Cause:
Inspection of all small bore stub joints of critical circuits was not done during shutdown.

Lesson(s):

- During plant piping inspection, all small bore stub joints of critical circuits should be covered to identify weak joints in a systematic way. Records must be maintained.

17. **Incident: Fire in offsite pipe rack due to a leak.**
Description:
Fire in offsite pipe rack between FOB (Fuel Oil Block) and TPS (Thermal Power Station) broke out due to a leak in the product line followed by auto-ignition in the hot surface.

Cause:
Periodic inspection of offsite pipelines was not done.

Lesson(s):

- Periodic inspection and repairs of offsite lines must be ensured through a proper schedule. Records of compliance and repairs must be maintained.

18. **Incident: Fire at CDU fractionator's bottom pump.**
Description:
Fire occurred at bottom pump due to failure of high point vent (HPV)/pressure gauge tapping followed by auto-ignition.

Cause:
Leak occurred due to shearing of HPV/pressure gauge tapping of the bottom pump.

Lesson(s):

- Provision to isolate the suction header at the column should be made for critical service pumps, such as those involving fluid at auto-ignition temperature. This will help early containment of fire at the bottom pump.

- For pumps at critical service, field visits/rounds must be ensured for early identification of abnormal conditions like high vibration.
- Health monitoring of connected joints like HPV, LPD (low point drain) should also be included in the NDT (non destructive test) of piping in critical service.

19. **Incident: Furnace damaged due to pressurization.**

Description:

In Paraxylene project during dry out operation, furnace box got pressurized and side walls were opened up by explosion, damaging the furnace. Two persons were injured.

Cause:

Furnace box purging was not ensured during burner light up.

Lesson(s):

- Furnace operating guidelines, including lighting up of first burner must be strictly followed and supervised.

20. **Incident: Fire in FCC fractionators column:**

Description:

During FCC start fire broke out in a FCC fractionators column after a leak developed due to sudden pressurization in the column when condensate (water) entered with steam used for dechoking purposes.

Cause:

Management of change procedure was not followed with respect to additional provision made for this purpose. The precautions and awareness for condensate draining prior to introduction of steam used for dechoking purpose were not followed.

Lesson(s):

- Any additional provision like introduction of steam must be duly approved with HAZOP recommendations complied. The change of management must ensure procedures and precautions required (like water draining from steam line prior to introducing steam inside columns) and awareness of the same amongst the concerned operators.

21. **Incident: Flash fire by grinding spark.**

Description:

At CDU in chemical dosing area, a flash fire of hydrocarbon vapors occurred. Contractor supervisor succumbed to burn injuries.

Cause:

Grinding spark provided ignition source.

Lesson(s):

- Adequate supervision must be ensured for proper co-ordination of jobs to contain grinding sparks and ensuring sparks do not reach the hazardous area. A hydrocarbon free atmosphere must be ensured by confirming by a gas test.

22. **Incident: Fire at a crude tank during sludge removal.**

Description:

Fire occurred at crude tank, 1c from which sludge was being removed through its bottom drop out door while preparing it for maintenance. The fire was kept confined to the tank with various firefighting measures. Nobody was injured.

Cause:

Hydrocarbon vapors were present inside the tank during sludge removal. Metal-to-metal contact caused a spark for ignition while pushing the drain pipe using scaffolding pipe.

Lesson(s):

- Frequent gas test should be done and ventilation should be continued during the tank cleaning process to ensure safe working conditions inside the tank.
- Critical activities such as cutting/removing of pipe and structural work should be supervised for adherence to safety conditions.
- During tank cleaning operation, a safety system must be available. Foam pourer also should remain in place to take care of emergencies.

23. **Incident: Auto ignition of leaked hydraulic oil in FCC.**

Description:

While cleaning the hydraulic oil filter of a spent catalyst slide valve system in FCC unit as per preventive maintenance checks, some hydraulic oil slipped and fell on the hot surface of a regenerator shell and caught fire. The unit was shut down as a precautionary measure.

Cause:
Auto ignition of spilled oil on a hot surface occurred.

Lesson(s):

- Ensure all precautions during maintenance of slide valves system to contain spillage and to prevent it coming in contact with any hot surface.

24. **Incident: Fire in TPS (Thermal Power Station) due to LDO (Light Diesel Oil) line rupture.**
Description:
During startup of Boiler-4, as LDO pump was started, a major fire broke out between TPS control room and boiler-3 firing floor.

Cause:
LDO supply line to boiler-4 ruptured due to corrosion, wrong line up with LDO screw pump operating without a pressure relief valve in system.

Lesson(s):

- Provision of pressure relief valve is to be ensured.
- Pipelines in hydrocarbon should not be encased with any other system.
- Suitability of pipe schedule and metallurgy is to be ensured.
- Cable entry points of TPS control room to have fire stop/fire barriers.
- Control fire suppression system should be provided.

25. **Incident: Fire in Hydrocracker due to exchanger flange leak.**
Description:
Tripping of recycle gas compressor (RGC) and system depressurization caused a leak from a flange joint in feed/reactor effluent exchanger and there was subsequent fire for 20 minutes. There was neither injury nor any damage to equipment.

Cause:
Thermal shock due to process fluctuations caused leak in exchanger.

Lesson(s):

- Ensure hot bolting of critical exchangers.
- Take actions to avoid RGC tripping.

Note: HAZOP review also should be done with respect to thermal shock and leak.

26. **Incident: Buoy of dismantled floating roof exploded.**
Description:
During gas cutting of dismantled places of old floating roof tanks within its dyke, one floating buoy exploded. It injured four contract workmen working at the site. Later at the hospital, two of them succumbed to injuries.

Cause:
The buoy which contained hydrocarbon was not made hydrocarbon free before initiating gas cutting on it.

Lesson(s):
- Analysis of possible hazards involved in a job must be done prior to undertaking the job. Accordingly, procedure for safe execution of the job must be prepared and implemented under strict supervision and guidance.
- Buoys should be cut only after removing the flange covers and filling them with water.
- In case the bolts of flanges are jammed, those should be cold cut for removal.

27. **Incident: Fire in an empty tank during a hot job with nearby pipeline rupture.**
Description:
In one refinery, a hot job was going on in one HSD tank. The tank was fully isolated from all connected lines, but one HSD transfer line passing near the tank was charged to start product transfer operation to the marketing tank. Suddenly, the pipeline started vibrating and causing a knocking sound; without giving scope to inform those concerned, the pipeline got ruptured at the location near the tank. The laborers inside the tank were drenched with spill of HSD and within no time fire started, resulting in human fatality.

Cause:
The ruptured pipeline found adequate pitting corrosion in the zone of rupture; from the records it was found that although piping replacement was carried out in other sections of the pipe that zone was also earmarked for replacement but was pending. This lack of co-ordination, namely checking all documents before giving a hot job permit, resulted into fire.

Lesson(s):

- While giving a hot job permit in a hydrocarbon free tank or any equipment or in any line, the hazard possibility surrounding the location of the hot job should be looked into; for example, surrounding pipelines/equipment operation is either to be stopped or their health (like inspection history or any fitting failure possibility in the charged line/equipment) should be critically looked into before giving out a hot job permit.
- There should be TSV (relief valve out of temperature rise) connected to long length pipeline with open end of TSV to be connected to a operating tank of same service.

28. **Incident: Fire in MS (Gasoline) tank while emptying out for maintenance.**

Description:

One MS tank content was pumped out for emptying for maintenance after the tank level became unpumpable, i.e., it went below the tank bottom manhole, water filling was done to raise the level and pumping was restarted. When the tank content became sufficiently diluted with water, the tank bottom content drained through an open funnel to ETP. Then the tank bottom manhole was opened. Simultaneously, disconnecting and lifting of the line connected MOV was started for repair/replacement. While doing so, there was a sudden explosion and fire which caused fatality also.

Cause:

When the tank bottom manhole was opened, MS—being very light—caused a vapor cloud in the zone. Before gas testing for an explosiveness check, a permit was given for lifting the MOV with a chain pulley. In the lifting mechanism, a non flame proof fitting was used, as usual. Hence, it might have caused a spark to connect with the MS vapor cloud and an explosion resulted.

Lesson(s):

- Simultaneous jobs are to be avoided. If extremely urgent, proper hazard study with an approval system should be followed.
- Using tool(s) with non-flameproof features come under the hot job category and thus, before any hot job, explosive test should be carried out with an explosive meter. So, after tank bottom manhole opening is done by gas test, the area should be ensured gas free.

29. **Incident: Fire in VGO-HDT (Vacuum gas oil hydrotreater) unit heat exchanger line.**

Description:

The heat exchanger train was consisting of a series of heat exchangers for unit feed versus reactor effluent heat exchange. As per design, there was a bypass line of the exchanger train for the feed. During operation there was a sudden explosion and fire around the exchanger train. It was observed that the pipeline rupture occurred at the elbow in the pipeline near the inlet of the hottest exchanger of the series. The injury update, if any, was not known but there was a major production loss due to outage of the unit for long.

Cause:

The unit was not so old that the elbow became corroded; however, it was found that unit was being operated at a capacity greater than design. When such things occur, it is usually the elbow, the weakest part in the pipeline, which erodes fastest and causes failure. Over and above, it was found that the elbow support leg was fixed to the foundation and thus the line did not get a chance for lateral movement during thrust; in addition, the bypass line meets the reactor outlet first in the exchanger inlet line through a quill—the quill being located at a very short distance of about 40 inches from the elbow as compared to minimum allowable distance of 10 times of pipe diameter of about 12–16 inches. All these deviations caused the failure of the elbow, resulting in explosion and fire.

Lesson(s):

- Operation at a higher capacity than design should not be done without HAZOP study, as well as without the licensor's consent.
- The design failure such as in the case of the quill location and wrong elbow support might be due to a lack of co-ordination between engineering and execution. Hence, the operating crew should not only check compliance with respect to P&ID during pre-commissioning, but also compare all fittings/fixtures like quill, thermo well, PSV, etc. connections along with all top, bottom support jobs. As such, operation should make a checklist together with their engineering colleagues.

e. **Inhalation of toxic gas:**

About 10% of case studies are found in this category and are described below:

1. **Incident: Inhalation H2S gas.**
   Description:
   In SRU (Sulphur Recovery Unit) of DHDS unit, while attending a leak in sour gas total bypass line, a maintenance technician fainted in his work spot and was declared dead when taken to hospital.

   Cause:
   PPEs were not used; lack of supervision, training and awareness; H2S presence was not checked before starting the job.

   Lesson(s):
   - Proper safety precautions like gas test at job spot must be stipulated in work permit.
   - Proper supervision must be ensured against compliance of safety precautions and use of appropriate PPE.

2. **Incident: Exposure to toxic atmosphere in a confined space.**
   Description:
   Two contract workers and one casual labor fell unconscious inside a sluice pit while working with a chain pulley block on the sluice gate of a surge pond drain. Subsequently when taken to hospital, they were declared dead.

   Cause:
   It happened due to unauthorized entry of the workers without PPEs in the sluice pit (confined space) where toxic gas, H2s was present. Lack of awareness of the workers and supervisors on precautions required before entering a confined space.

   Lesson(s):
   - Based on job safety analysis, methodology for carrying out maintenance and operation jobs inside a confined space must conform to stipulations of OISD-105 with respect to work permits, gas tests, use of PPEs and proper supervision.
   - Safety awareness on presence of toxic gases like H2S inside a sluice pit should be created amongst all workmen and contract supervisors.

3. **Incident: Fatality due to inhalation of chemicals.**
   Description:
   While decanting DMDS into the system, the chemical splashed over the technician as its drum got pressurized and burst open.

Cause:

Leakage occurred due to over pressurization.

Lesson(s):

- Use of required PPEs to be ensured.
- The system for decantation should be made operational after HAZOP study.

4. **Incident: Fatality in confined space; namely Flare KOD (Knock out drum).**

   Description:

   During CRU turn around, a contractor worker expired in a flare knock out drum.

   Cause:

   It was due to oxygen deficiency in confined space or in presence of hydrocarbon vapor without sufficient ventilation and proper PPE.

   Lesson(s):

   - Prior to worker entry into a confined space, actual compliance of work permit conditions must be ensured in line with OISD-105.
   - Awareness level and adequate supervision must be ensured.

5. **Incident: Inhalation of naphtha.**

   Description:

   While attempting to dechoke the water drain line of a feed surge drum boot of NHTU, the line became suddenly dechoked, which resulted in injury to working personnel, along with inhalation of naphtha vapor. The worker succumbed to the injury.

   Cause:

   Lack of supervision and non-adherence of SOP.

   Lesson(s):

   - Jobs should be carried out with proper job safety analysis and with better supervision.

6. **Incident: Fatality due to slip into sewer.**

   Description:

   One contract worker slipped inside a sewer chamber while removing the trash from the sewer by using a mechanical hand operated grabber. He was seen inside the sewer manhole;

pulled out with a rope life line and rushed to the occupational health center, where he was declared dead.

Cause:

There was a lack of supervision.

Lesson(s):

- Better supervision required.
- On-the-job safety training to be strengthened.

7. **Incident: H_2S gas inhalation in DHDS unit.**

Description:

In one short shutdown, the sour gas knock out drum in DHDS unit was handed over to maintenance after isolation of the same to carry out a maintenance job inside the vessel. The maintenance engineer, before sending the laborer inside the vessel, put his head inside the vessel through its bottom manhole standing in the platform outside to check where there was any smell of H_2S or not. But the engineer became senseless and, not coming out, was taken to first aid. Following treatment he recovered and thus there was no fatality.

Cause:

The vessel, upon checking, was found to have all its nozzles not blinded. One nozzle remained unblinded, causing sour gas entry inside the vessel.

Lesson(s):

- In a work permit system; there should be a maker and a checker; if one person signs the work permit, the checker should not hand over the same to maintenance before checking it at site.

8. **Incident: Toxic gas inhalation in FCC unit.**

Description:

A sour gas knock out drum was being drained off condensate to an open sewage funnel by operation; as the liquid level was high, the operator left the place keeping the drain valve open. It was during lunchtime and one laborer, who was engaged in some other area, was sleeping in that zone unattended. When after some time, the operator came to close the drain valve, he found that the laborer was senseless. Upon taking the laborer to hospital, it was found that he had succumbed to H_2S inhalation.

Lesson(s):

- The operator should not leave the place having kept the drain valve open.

f. **Electrocution/Electric Flash:**

1. **Incident: Electrocution at ETP (Effluent treatment plant).**
   Description:
   A contract worker at ETP was electrocuted as he slipped and accidentally touched the limit switch without a top cover.

   Cause:
   Electrical protections not ensured. Handrails not provided.

   Lesson(s):

   - Electrical protections on the panels switches must be ensured through scheduled periodic checks, and repairs and records must be maintained.
   - Protection against slip/trip/fall also should be ensured.
   - Job site should be checked by the engineer in charge.

2. **Incident: Electrocution due to improper earthling connection.**
   Description:
   While lifting grinding machine, a contract worker fell on the ground and died.

   Cause:
   Electrocution caused fatality because the grinding machine was not securely earthed. Its cable was not connected with the pin of the plug.

   Lesson(s):

   - Earthling connection of the portable equipment needs to be checked and to be ensured by an electrician before starting the job.

3. **Incident: Flashover in switch gear panel.**
   Description:
   A maintenance technician suffered burn injury due to electrical flashover in 33KV switch gear breaker panel.

   Cause:
   Possibly the victim's right hand accidentally entered into the prohibited zone, crossing the arc flash protection boundary from the energized parts.

Lesson(s):

- Nobody should be allowed to work inside the cubicle when the bus is energized.
- Follow guidelines as per NFPA 70E for minimum insulation distances to avoid flashover.

4. **Incident: Fatality due to electric shock.**

   Description:

   Fatality of a contract laborer occurred due to electric shock, as the welding cable joint was touched during shifting of the welding machine.

   Cause:

   It was a case of lack of supervision and non-use of PPEs.

   Lesson(s):

   - Use of required PPEs and increased supervision is to be ensured.

g. **Poor workmanship/design:**

1. **Incident: Roof slab collapsed during roof casting.**

   Description:

   At a construction site of the Naphtha Cracker complex during roof casting of a satellite rack room, a portion of the slab with supporting structure collapsed upon 13 persons working underneath. One contract workman got trapped in the debris and died, and eleven workers received minor injuries.

   Cause:

   There were deficiencies in design, material inspection and erection supervision of the supporting system of the work.

   Lesson(s):

   - Design adequacy of temporary enabling work to be ensured.
   - Execution should be done only after the final approved drawing is available. Deviation, if any, should be documented.
   - For temporary but critical works a formal quality assurance plan should be approved by PMC (Project Management Consultant).
   - Safety supervision of enabling jobs is to be ensured.
   - Use of pre-fabricated, pre-engineered and intrinsically safe systems of work is to be promoted.
   - Pipes from approved vendors only must be used.

2. **Incident: Fuel tank roof caved in and collapsed.**
   Description:
   In Hydrocracker project, the fixed roof of a fuel oil tank caved in, along with a top course of the tank shell.

   Cause:
   Creation of a vacuum in the tank possibly due to an unintended high withdrawal rate of product from the tank caused the failure of the roof and supporting structure. The weakened roof subsequently collapsed.

   Lesson(s):
   - Pumping out rate is to be monitored for all connected tanks and limits should be made known to all operating personnel.
   - A pressure transmitter should be provided on one cone roof tank with an alarm on vacuum condition.
   - The breather vent of the tank should be periodically checked to see of a bird nest has formed or not.
   - On-off position of MOVs should be displayed in control room.
3. **Incident: Injury during maintenance activity.**
   Description:
   While deplugging the set concrete from the hopper top, the agitator was rotated and the contract operator's leg below the knee suffered multiple fractures, resulting in amputation of the leg. About 30,000 man-hours were lost in the incident.

   Cause:
   Work procedure was not followed.

   Lesson(s):
   - Jobs should be carried out with proper job safety analysis and with better supervision.
4. **Incident: Poor handling of material.**
   Description:
   While shifting project material like transporting one 16-inch diameter pipe by a hydra, the helper holding the pipe was hit by it during a jerk caused by the hydra during its movement.

   Cause:
   It was due to an incorrect way of transporting material—moving the pipe using hydra, instead of taking it inside a trailer truck.

Lesson:

- Safe procedures should be followed instead of short cuts, even though it was a single pipe movement.

5. **Incident: Fatality during earth excavation in residential colony.**

   Description:

   Earth excavation was started to construct a swimming pool inside the residential colony. During lunchtime, a report came that one laborer had expired in a landslide while excavating.

   Cause:

   The excavator and labor were not working in the presence of a supervisor; it was lean time also from human movement point of view; thus, the lack of care on the job was missed by the supervisor.

   Lesson(s):

   - When excavator is in operation, there should be no laborer inside the pit.
   - The job should be stopped when the supervisor leaves the location.
   - Laborers should not be allowed to work close to the wall of the pit because there is a high chance of landslide in those zones.

h. **Miscellaneous:**

1. **Incident: Fatality during the cleaning of an open stormwater channel.**

   Description:

   A contract worker, while cleaning, was found immobilized and lying in the channel near the POL wagon gantry. When he was taken to hospital, he was declared dead.

   Cause:

   Job analysis safety was not carried out.

   Lesson(s):

   - Better supervision and alertness required with prior job safety analysis.

2. **Incident: Missing and fatality with unknown reason.**

   Description:

   During housekeeping at a process unit of refinery, one contract worker was found missing for two days and eventually found lying dead near a cooling tower.

Cause:

Lack of supervision control and administration.

Lesson(s):

- Better supervision and administration control for workers; i.e. keeping track of their wherabouts, must

9.2.2 Case Studies Outside India

1. **Incident: Fire by lie rupture in HCU (Hydrocracker Unit).**

   Description:

   A Chevron refinery, Richmond, USA suffered an accident in 2013 due to a line rupture in HCU resulting in fire. While the injury/fatality figure is not known, the unit suffered property damage, as per news from CSB (Chemical Safety and Investigation Board), USA. The accident occurred due to the sudden rupture of a pipeline in HCU during operation.

   Cause:

   It was said that rupture was due to sulfidation corrosion in the pipeline.

   Author's Analysis:

   In any hydrotreatment unit, sulfur is converted into hydrogen sulfide in the reactor of the unit. Hence, all the piping coming out from the reactor outlet contains hydrogen sulfide until it is separated out in the downstream equipment. Any equipment falling in this zone would also contain hydrogen sulfide. In the reactor, ammonia is also formed due to the presence of nitrogen in the feed. Hydrogen sulfide reacts with ammonia to form ammonium salt of hydrogen sulfide, which is a solid product and deposits in the pipeline and/or equipment if this salt is not properly washed away by water where it is soluble. As ammonium salt is an unstable compound, properties of hydrogen sulfide remain intact; hence, hydrogen sulfide starts attacking the carbon steel and causes stress corrosion cracking of the pipe and equipment; this is also called sulfidation corrosion. Due to this reason, industries have introduced water washing of the piping and equipment falling at the downstream of the reactor. The washed away ammonium salt of hydrogen sulfide is separated in a vessel from where the washed water called sour water is taken out from the bottom of the vessel under level control. This sour water subsequently goes to another unit called the sour water stripper unit to strip out the hydrogen sulfide, and the hydrogen sulfide then goes to the sulfur recovery unit to recover sulfur from it. All metallurgical care is taken in the sour water line and downstream unit, as mentioned above.

Now, as discussed above, in spite of having a wash water facility in downstream of reactor in HCU, if corrosion still takes place, then it must be either any deficiency in the wash water injection facility design or any fault in the operation of watering up to a desired level and up to a desired period in a working shift, as decided by the expert.

Lesson(s):

- The location of the quill for injection of wash water, quill design, wash water flow distribution in all concerned areas, wash water flow rate, continuity and uninterrupted injection in a working shift should be reviewed to identify the deficiency.
- If one industry thinks that there may be any deficiency in the washing system, in that case, they should strengthen the metallurgy of the piping and equipment falling in that zone which currently is followed by the industries.

2. **Incident: Fire in DCU (Delayed Coker Unit).**
Description:
Anacortes refinery, Washington, USA suffered a fire in 1998 at the bottom of a coke drum in their DCU, as per news report issued by HistoryLink.org (dated 25 November, 1998). As per report, the unit underwent a forced shutdown due to a power outage. After power was restored, they initiated unit restart activities which took place about two days after shutdown but encountered some problems. As a part of the identification activities, they realized that the coke drum bottom inlet line got choked. Then they tried to start a deheading operation at the coke drum bottom which is an activity after completion of the coking cycle in the drum followed by water cooling, as per facility provided. In doing so, suddenly, some hot liquid along with coke came out and caught fire killing some people at work by burning.

Cause:
The auto ignition temperature of a DCU feed is very low and so would be that level of unconverted liquid, if it remained in the drum. If the temperature of the drum material is not adequately cooled, then residual liquid—if it remains—would be of high enough temperature to cause auto ignition when it comes in contact with air, which must be the reason for the abovementioned fire.
Author's analysis:
After any emergency shutdown including power interruption, the heater coils of DCU is to be thoroughly flushed up to the coker drum inlet immediately after failure with steam header pressure available. Also, utility boiler of the refinery is to be immediately started with power from DG set if the main power supply is not

available. In this steam flushing, heater coils along with the heater's outlet line up to the coker drum inlet get flushed and there remains no chance of choking of the lines. Also, when all power and utility is restored, the coke drum water cooling should be started without depending on any temperature indication that coke drum is sufficiently cooled during long idling. It may be mentioned that no temperature instrument readings are reliable during idling of operation. Also, as thermal conductivity of a liquid is very low compared to a solid, cooling is done using the heat convection property of liquid, i.e., using water cooling in this case. If one calculates to find out the time required for cooling by idling instead of water cooling, it would be found that it takes an enormous time for cooling. Operation after doing this calculation is an unfeasible task: that's why the operating crew should follow the laid down procedures. That could be the cause of the problem that occurred in the above accident.

Lesson(s):

- The operating manual should be reviewed and amended if any inadequacy is found.
- The operation should not violate the standard operating procedures.

3. **Incident: Fire in raffinate splitter of Isobutane ISOM Unit.**
Description:
BP Texas city refinery suffered a major fire in 2005, as per news report by Wikipedia.org dated 23rd March, 2005.
The objective of the raffinate splitter is to separate heavy distillate in the ISOM unit. The process flow scheme is similar to that in sub sl. no. 2 of section d of main section "A" above.
Like in that case, here there was also a blow down vessel to accommodate safety valve release material, if needed. But the vessel overflowed and the spilled over liquid had some vapor in it due to its hot condition. The vapor caught fire due to coming in contact with some ignition source nearby. It is also report that the fire caused human fatality and injury.

Cause:
It is learned that during startup while filling up the splitter, liquid is taken at a sufficient level to avoid emptying out of the splitter column during startup circulation as is done usually in startup. But level indication of the splitter might have been misleading, causing overfilling of the splitter and resulting in its safety valve popping and releasing the hot liquid to the blow down vessel; the liquid after being over flown from the top of the blow down vessel spilled to the

ground in the surrounding area. Fire might have occurred due to spilled vapor coming in contact with an ignition source nearby.
Author's analysis:
Though it is heavy distillate, it may contain some light distillate as normally happens; the liquid is light enough to create some vapor in a hot condition resulting in an explosion and fire coming in contact with the ignition source, if any.

Lesson(s):

- There should be sufficient auto cooling by a water flow mechanism in a blow down vessel, as mentioned in earlier case.
- There should two level indications to cross-check the reliability of one indication with respect to other.

4. **Incident: Fire at Philadelphia Energy Solutions Refinery Complex.**
Description:
Fire broke out on June 21, 2019, as per CNBC breaking news. Propane along with the chemical hydrofluoric acid was flowing through a pipeline. Suddenly, the pipe ruptured—most likely at a faulty elbow, as reported by CSB (Chemical Safety Investigation Board), USA. There were injuries to about five persons as per REUTER news.

Cause:
This was due to a failure of a corroded line in a age-old refinery.
Author's Analysis and Lesson(s):

- Quality assurance of equipment, piping and engineering is of the utmost importance. Irrespective of whether it is an old or middling or new installation, there should be a standard inspection and subsequent repair/replacement jobs planned and implemented for operating a plant without failure. In old lines, under insulation corrosion and external pitting in pipelines, internal erosion due to operating at excess capacity—particularly at elbows—are very common. Hence, a rigorous inspection program is essential.

9.3 Accident Case Studies in Gas Processing Plants

Some case studies in India could be traced out. These are discussed as below:

1. **Incident: Fire in storm water channel.**
Description:
Near the time of a shift change a fresh (spare) charcoal filter located in a gas sweetening unit (GSU) was taken in line and the spent one was

isolated. The charcoal filters are used for treating naphtha stream for color improvement. Soon the naphtha spill started from the back wash outlet valve of the fresh filter and migrated into the storm water channel crossing the boundary wall. Fire broke outside the boundary wall and quickly traversed through the channel towards the source of the hydrocarbon spill inside the plant. A major fire was effectively brought under control in about ten minutes, and its entry to the process unit was avoided. Two outsiders present near the storm water channel received burn injuries and one of them succumbed to their burn injury.

Cause:

The backwash outlet valve of filter was not closed tight and an end blind provision was not used for positive isolation, as is normal practice.

Delayed action on LEL alarm from the gas detector was another reason.

An effluent pit at the unit battery limit was kept lined up to storm water channel instead of ETP.

There was no oil catcher in the storm water channel.

Lesson(s):

- Approved procedures for filter changeover should be incorporated in the operating manual, stressing positive isolation using an end blind at the back wash drain valve. The blind should be removed only after regeneration of the charcoal bed.
- Standing instructions should be issued to carry out field checks wherever the gas detector goes on alarm mode.
- As far as possible, filter changeover is to be avoided during shift changeover.
- An oil catcher should be provided in the storm water channel to capture any accidental spill/leak.
- Back wash drain outlet should be routed to a closed system.

2. **Incident: Fire due to loss of containment during shutdown.**

Description:

During shutdown of condensate fractionation unit-II, a fire broke out on release of hydrocarbon vapors from an open flange in a pump suction header. The proximity of a hot work site provided an ignition source for the flash fire and one fatality occurred.

Cause:

Positive isolation of circuits was not ensured; no circuit was made hydrocarbon free.

Lesson(s):

- While preparing for a hot job during shutdown maintenance of the plant, positive isolation must be ensured.

3. **Incident: Fire at cooling tower at LPG area.**

   Description:

   In a LPG area, the fire took place in a cooling tower FRP hood causing damage to cables and the deck floor of the cooling tower.

   Cause:

   Fire occurred possibly due to a spark caused by short-circuit in the power/control cables in hydrocarbon atmosphere. Gas detectors were not in place and the presence of hydrocarbon was thus unnoticed.

   Lesson(s):

   - Gas detectors need to be installed and made functional.
   - The grounding system of the cooling tower pipeline network needs to be reassessed for proper earthing and corrective measures taken.

4. **Incident: Explosion and fire by rupture of natural gas (NG) line in MDPE plant.**

   Description:

   One road crossbridge over a natural gas pipeline was supposed to be constructed. For the purpose, the gas line going to MDPE (Medium density Polyethylene) was decommissioned, isolated and evacuated. Then excavation was started, but unfortunately, JCB hit another gas line passing nearby and leading to the other plant in operation, unknown to them. This caused the rupture of the live gas line, resulting in explosion and fire.

   Cause:

   The record of underground piping layout was not explored before issuing an excavation permit. The job execution permit without home work caused the pipe rupture.

   Lesson(s):

   - Before issuing a permit for excavation, underground piping documents must be reviewed.
   - A supervisor should be present at the site while the contract workers are executing the job.

References in Consequence Modeling

1. World bank technical paper no. 55 on techniques for assessing industrial hazards, 1988.
2. G. Madhu, Modeling of pool fires, *Journal of Fire Engg.*, 1 (2002), 1–10.
3. G. Heskestad, Engineering relations for fire planes, *Fire Safety Journal*, volume 7, issue 1(1984), p25–32.
4. Q. A. Baker, P. A. Cox, P. S. Westive, J. J. Kulese, *Explosion hazards and evaluation*, Elsevier, New York, (1983).
5. R. W. Prug, *Quantitative evaluation of BLEVE hazards, 22nd Loss prevention symposium*, AIChE, New York, (1988).
6. A. F. Roberts, Thermal radiation hazards from release of LPG from pressurized storage, *Fire Safety Journal*, (1981), p197–212.
7. Hymes, The physiological and pathological effects of thermal radiation, SRD R 275, Atomic Energy Regulatory Authority, UK, (1981).
8. AIChE/CCPS, *Guidelines for chemical process quantitative risk analysis*, 2nd edition, New York, (2000).
9. Pieterson, Huerta, Analysis of LPG incident in San Juan Ixhuattepec, Mexico City, TNO 84-022, *Netherland organization for applied scientific researches*, (1985).
10. TNO, methods for calculation of physical effects of the escape of dangerous materials: Liquids and gases (The yellow book), Netherland organization for applied scientific researches, Apeldoorn, the Netherland, (1979).
11. Brown, energy release protection for pressurized system, part I, review of studies into blast and fragmentation, *Applied Mechanics, Rev* 38(12), (1985), p1625–1651.

Index

Note: *Italicized* page numbers refer to figures, **bold** page numbers refer to tables.

For Product Safety Concerns and Information please contact our EU representative GPSR@taylorandfrancis.com Taylor & Francis Verlag GmbH, Kaufingerstraße 24, 80331 München, Germany

Batch number: 10725802

Printed by Printforce, the Netherlands